Introduction
to
Quantum Optics

Documents on Modern Physics

Edited by

ELLIOT W. MONTROLL, *University of Rochester*
GEORGE H. VINEYARD, *Brookhaven National Laboratory*
MAURICE LÉVY, *Université de Paris*

A. ABRAGAM   L'Effet Mössbauer

S. T. BELYAEV   Collective Excitations in Nuclei

P. G. BERGMANN and A. YASPAN   Physics of Sound in the Sea: Part I Transmission

T. A. BRODY   Symbol-manipulation Techniques for Physics

K. G. BUDDEN   Lectures on Magnetoionic Theory

J. W. CHAMBERLAIN   Motion of Charged Particles in the Earth's Magnetic Field

S. CHAPMAN   Solar Plasma, Geomagnetism, and Aurora

H.-Y. CHIU   Neutrino Astrophysics

A. H. COTTRELL   Theory of Crystal Dislocations

J. DANON   Lectures on the Mössbauer Effect

B. S. DEWITT   Dynamical Theory of Groups and Fields

R. H. DICKE   The Theoretical Significance of Experimental Relativity

H. FESHBACH and F. S. LEVIN   Reaction Dynamics

P. FONG   Statistical Theory of Nuclear Fission

E. GERJUOY, B. YASPAN and J. K. MAJOR   Physics of Sound in the Sea: Parts II and III   Reverberation, and Reflection of Sound from Submarines and Surface Vessels

M. GOURDIN   Lagrangian Formalism and Symmetry Laws

D. HESTENES   Space–Time Algebra

J. G. KIRKWOOD   Dielectrics—Intermolecular   Forces—Optical   Rotation

J. G. KIRKWOOD   Macromolecules

J. G. KIRKWOOD   Proteins

J. G. KIRKWOOD   Quantum Statistics and Cooperative Phenomena

J. G. KIRKWOOD   Selected Topics in Statistical Mechanics

J. G. KIRKWOOD   Shock and Detonation Waves

J. G. KIRKWOOD   Theory of Liquids

J. G. KIRKWOOD Theory of Solutions

V. KOURGANOFF Introduction to the General Theory of Particle Transfer

R. LATTÈS Mcthods of Resolution for Selected Boundary Problems in Mathematical Physics

B. W. LEE Chiral Dynamics

J. LEQUEUX Structure and Evolution of Galaxies

J. L. LOPES Lectures on Symmetries

F. E. LOW Symmetries and Elementary Particles

A. MARTIN and F. CHEUNG Analyticity Properties and Bounds of the Scattering Amplitudes

P. H. E. MEIJER Quantum Statistical Mechanics

M. MOSHINSKY Group Theory and the Many-body Problem

M. MOSHINSKY The Harmonic Oscillator in Modern Physics: From Atoms to Quarks

M. NIKOLIĆ Analysis of Scattering and Decay

M. NIKOLIĆ Kinematics and Multiparticle Systems

H. M. NUSSENZVEIG Introduction to Quantum Optics

J. R. OPPENHEIMER Lectures on Electrodynamics

A. B. PIPPARD The Dynamics of Conduction Electrons

H. REEVES Stellar Evolution and Nucleosynthesis

L. SCHWARTZ Application of Distributions to the Theory of Elementary Particles in Quantum Mechanics

J. SCHWINGER Particles and Sources

J. SCHWINGER and D. S. SAXON Discontinuities in Waveguides

M. TINKHAM Superconductivity

J. VANIER Basic Theory of Lasers and Masers

R. WILDT Physics of Sound in the Sea: Part IV Acoustic Properties of Wakes

Introduction
to
Quantum Optics

H. M. NUSSENZVEIG

Institute for Fundamental Studies
Department of Physics and Astronomy
University of Rochester
Rochester, New York

GORDON AND BREACH SCIENCE PUBLISHERS

London   New York   Paris

Editorial office for the United Kingdom

Gordon and Breach, Science Publishers Ltd.
42 William IV Street
London W.C. 2.

Editorial office for France

Gordon & Breach
7–9 rue Emile Dubois
Paris 14ᵉ

To Micheline

Preface

THIS BOOK is based on lectures given at the Latin-American School of Physics held in La Plata in 1970. The actual lectures covered the material in Chapters 1 through 6 and part of Chapter 7. The remainder of Chapter 7 and Chapter 8 were added later. Finally, Appendix A has been added to cover some recent developments in the rapidly expanding field of coherent atomic interactions.

The lectures were aimed at theoretical physicists and graduate students. They were meant as an introduction to a new subject, with which no previous familiarity was assumed. Many of the topics have been reviewed before in some excellent articles and sets of lectures; I have drawn considerably from these sources. Some new results have also been included.

My own interest in and knowledge of the subject have benefited greatly from my interaction with the quantum optics group at the University of Rochester. I would like to thank, in particular, Professors J. H. Eberly, L. Mandel and E. Wolf. I would also like to thank Mrs. Shirley G. McDonnell for the very skillful typing of the manuscript. The partial support of a grant from the Advanced Research Projects Agency is gratefully acknowledged.

H. MOYSÉS NUSSENZVEIG

10355

Contents

Preface ix

1 Classical Coherence Functions **1**

   1.1 Introduction 1

   1.2 Temporal and spatial coherence 1

   1.3 Classical theory of coherence 4
      (a) Analytic signals 4
      (b) Second-order coherence functions 6
      (c) Interference spectroscopy and stellar interferometry 8
      (d) Higher-order coherence functions 12

2 Quantum Coherence Functions **17**

   2.1 The quantized radiation field 17

   2.2 Theory of photoelectric detection 20
      (a) Introduction 20
      (b) Single-atom detector 22
      (c) Higher-order coherence functions 25
      (d) Emission counters 27

   2.3 General properties of coherence functions 28

   2.4 Coherent fields 34
      (a) Second-order coherence 34
      (b) Higher-order coherence 36
      (c) Density operators for coherent fields 38

3 Coherent States **41**

   3.1 Definition of coherent states 41
      (a) Introduction 41
      (b) Coherent states for a single mode 42

   3.2 Properties of coherent states 45
      (a) Coherent states as displaced harmonic oscillator
         states 45

(b) Coherent states as minimum-uncertainty states 47
(c) Coherent states and the forced harmonic oscillator 48
(d) Non-orthogonality and overcompleteness 49
3.3 Representations in terms of coherent states 50
(a) Representation of state vectors 50
(b) Representation of operators 53
3.4 The diagonal representation 54
(a) Introduction 54
(b) The optical "equivalence" theorem 56
(c) The characteristic functions 62
(d) Extension to several modes 64
(e) Discussion 66

4 Applications **71**

4.1 Examples of light fields 71
(a) Thermal light 71
(b) Ideal laser light 73
4.2 Photon counting probabilities 75
(a) The counting distribution 75
(b) Examples 76
4.3 Higher-order coherence effects 78
(a) Coherence functions for Gaussian light 78
(b) Bunching effects 80
(c) The Hanbury Brown and Twiss effect 81
(d) Interference of independent beams 82
4.4 The reconstruction theorem 84

5 The Laser **89**

5.1 Introduction 89
5.2 Optical resonator modes 91
5.3 Elementary theory of the threshold 94

6 Semiclassical Theory of the Laser **99**

6.1 Introduction 99
6.2 The field equations 102
(a) The frequency and amplitude equations 102
(b) Examples 105

6.3   The microscopic polarization   107
      (a)   Schrödinger equation for a 2-level system   107
      (b)   Geometrical interpretation   111

6.4   The macroscopic polarization   113

6.5   The linear approximation   117

6.6   Nonlinear theory   121
      (a)   Steady-state solution of the atomic equations   121
      (b)   The field equations   124

6.7   Mode competition   126

6.8   Effect of atomic motion   132

7   **Quantum Theory of the Laser** **137**

7.1   Introduction   137

7.2   The reduced density operator   139

7.3   Laser model and Hamiltonian   141
      (a)   The laser model   141
      (b)   Field quantization   143
      (c)   Hamiltonian   144

7.4   Equations of motion   145
      (a)   The combined system   145
      (b)   Perturbation solution   146
      (c)   Reduced density operator for the field   148

7.5   Photon statistics   151

7.6   Transient behavior and linewidth   156
      (a)   Transient buildup from vacuum   156
      (b)   The intrinsic linewidth   157
      (c)   Physical interpretation   160

7.7   Other approaches   161
      (a)   The Fokker-Planck equation   161
      (b)   The Langevin method   166
      (c)   The quantum regression theorem   168

8   **Superradiance, Photon Echoes and Self-induced Transparency** . **171**

8.1   Introduction   171

8.2   Properties of two-level systems   172

8.3   The cooperation number   174

8.4 Superradiance   177
   (a)   Superradiant states   177
   (b)   Classical model   179
   (c)   Extension to other cases   182

8.5 Photon echoes   183

8.6 Self-induced transparency   187
   (a)   The McCall-Hahn equations   187
   (b)   The hyperbolic-secant solution   191
   (c)   The area theorem   196
   (d)   Discussion   200

8.7 Concluding remarks   203

Appendix   Recent Developments in Quantum Optics   .   .   . **207**

A.1 Survey of some recent experimental results   207

A.2 Coherent atomic states   209

A.3 Field quantization in Dicke's model   218

A.4 Stationary states   220

A.5 The thermodynamic limit   224

A.6 Dynamics of superradiant emission   231

Index   239

CHAPTER 1

Classical Coherence Functions

1.1 INTRODUCTION

IMPORTANT NEW developments have taken place in optics during the past decade. They have been made possible mainly by the availability of a new type of light source, the laser. Improvements in detection techniques, allowing one to detect very weak and very fast signals, have also played an important role.

These advances have led to the exploration of new domains, like that of nonlinear optics, to the discovery of new physical effects, like self-induced transparency, and to the investigation of some very fine structural properties of the electromagnetic field and of its interaction with matter.

From the theoretical point of view, we have learned how to characterize the statistical properties of the field in terms of the quantum theory of coherence. A new representation for the field, in terms of so-called coherent states, has been found, and it appears to have many important applications. The laser itself is an inherently nonlinear device, as well as an example of an open system very far from thermal equilibrium, and the treatment of its operation raises many interesting theoretical questions. The nonlinear effects arising in the interaction of laser light with matter have led to several surprises and beautiful new phenomena.

The basic tools for dealing with these problems are quantum electrodynamics (semiclassical approximations are often useful) and statistical mechanics. The whole field is usually referred to as *quantum optics*.

The present lectures are intended as an introduction to this very broad and rapidly expanding subject. Only some of the main topics will be treated. Considerably more detailed treatments may be found in several books[1-4] and other references that will be given[5].

1.2 TEMPORAL AND SPATIAL COHERENCE

A good deal of classical optics is concerned with the propagation of plane monochromatic waves. In such a wave there is always a definite phase relation between the fields at two different points in space or time, no matter how far apart they may be. Such phase relations can be manifested, in principle, by bringing together different portions of the field and observing the resulting interference effects.

Plane monochromatic waves, however, are just a convenient mathematical fiction. Real light sources consist of large numbers of atoms; the emission from different atoms may be completely independent or it may be correlated to some extent, but both for the light emitted by a given atom as well as for emissions from different atoms, the relative phase relations are subject to random fluctuations, arising from thermal motion and many other effects. Thus, if we sample the field at two different points in space or time, no definite phase relations are found for large enough separation.

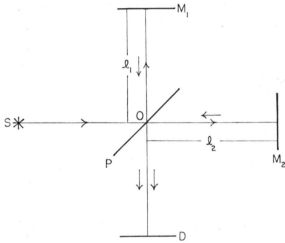

Figure 1.1   Michelson interferometer. S: light source; P: semitransparent plate; M_1, M_2:
mirrors; D: detector

The Michelson interferometer (Fig. 1.1) allows us to compare phase relations at two different times by splitting the light beam at the point 0 and reuniting the beams with a time delay $\tau = (l_2 - l_1)/c$ between the components. The observed interference fringes are a measure of the correlation between the field at the point 0 at a time t and the field at the same point at the time $t + \tau$.

As τ increases, the sharpness of the interference fringes usually decreases, until they become completely washed out, for τ much larger than a characteristic time $\Delta\tau$, called the *coherence time*. For an ideal, perfectly monochromatic beam, $\Delta\tau$ would be infinite; for a real light beam, having a spectral frequency distribution of width $\Delta\nu$, we expect that

$$\Delta\tau \sim \frac{1}{\Delta\nu}. \tag{1.2.1}$$

The length $\Delta l = c\Delta\tau$ is called the *coherence length*. For the most highly monochromatic thermal light sources, $\Delta\tau$ is at most of the order of the natural lifetime of an excited atomic state, $\Delta\tau \lesssim 10^{-8}$ sec, corresponding to $\Delta l \lesssim 10^2$ cm. For a laser, $\Delta\tau$ can exceed 10^{-7} sec, with $\Delta l \gtrsim 10^8$ cm

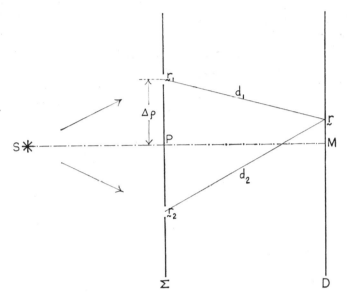

Figure 1.2   Young's experiment. S: light source; Σ: screen; r_1, r_2: pinholes; D: detector

The familiar Young interference experiment (Fig. 1.2) allows us to compare phase relations between two different points r_1, r_2 in the field. If the distance $2\Delta\varrho$ between the pinholes at r_1, r_2 is too large, the interference fringes observed at r disappear. We can thus define a *coherence area* $\Delta A \sim (\Delta\varrho)^2$ around a point P in the field.

The coherence area and the coherence length can be combined into the *coherence volume* $\Delta V = \Delta A \cdot \Delta l$. The coherence volume around a point P corresponds roughly to the volume of the field from which we may draw samples capable of producing interference with the field at P.

It is not difficult to show[6] that the coherence volume corresponds to one cell in the quantum-mechanical phase space associated with a photon, so that the above interpretation is consistent with what one would expect in quantum mechanics: interference is possible when we are dealing with a single quantum state.

The average number of photons in the same state of polarization contained within a coherence volume therefore represents their average number in a given quantum state, which is known, in accordance with quantum

1*

statistical mechanics, as the *degeneracy parameter* δ of the radiation[7]. For blackbody radiation, this is given by the well-known Einstein formula

$$\delta = \frac{1}{e^{h\nu/\varkappa T} - 1}. \tag{1.2.2}$$

For thermal light sources, $\delta \lesssim 10^{-3}$, so that thermal light is nondegenerate. On the other hand, laser light is typically highly degenerate ($\delta \gg 1$); values of δ exceeding 10^{14} have been obtained. This leads to fundamental differences between the statistical properties of thermal light and those of laser light, as will be seen later.

1.3   CLASSICAL THEORY OF COHERENCE

(a)   Analytic Signals

In order to simplify the notation, let us replace the electromagnetic field by a scalar field $V^{(r)}(\mathbf{r}, t)$ (this can be thought of, for instance, as one component of the vector potential or of the electric field). Polarization effects can be separately dealt with[8].

Let us consider the Fourier decomposition of $V^{(r)}(\mathbf{r}, t)$,

$$V^{(r)}(\mathbf{r}, t) = \int_{-\infty}^{\infty} v(\mathbf{r}, \nu)\, e^{-2\pi i \nu t}\, d\nu. \tag{1.3.1}$$

Since $V^{(r)}$ is real, we have

$$v(\mathbf{r}, -\nu) = v^*(\mathbf{r}, \nu). \tag{1.3.2}$$

Thus, the positive-frequency components already contain all the information, and we may consider, instead of $V^{(r)}$, the associated *analytic signal*[9], defined by

$$V(\mathbf{r}, t) = \int_{0}^{\infty} v(\mathbf{r}, \nu)\, e^{-2\pi i \nu t}\, d\nu. \tag{1.3.3}$$

Clearly,

$$V^{(r)}(\mathbf{r}, t) = 2\,\mathrm{Re}\, V(\mathbf{r}, t). \tag{1.3.4}$$

The reason for the name "analytic signal" is that $V(\mathbf{r}, t)$ can be analytically continued into the lower half of the complex t-plane, $\mathrm{Im}\, t < 0$, as may readily be seen by substituting t by $t + i\tau$ in (1.3.3). If the instantaneous intensity integrated over all times is finite, i.e., if [cf. (1.3.7)]

$$\int_{-\infty}^{\infty} |V(\mathbf{r}, t)|^2\, dt < \infty, \tag{1.3.5}$$

one can apply Titchmarsh's theorem[10] and conclude that Re $V(\mathbf{r}, t)$ and Im $V(\mathbf{r}, t)$ are Hilbert transforms of each other:

$$\text{Im } V(\mathbf{r},t) = \frac{1}{\pi} P \int_{-\infty}^{\infty} \frac{\text{Re } V(\mathbf{r}, t')}{t' - t} dt',$$

(1.3.6)

$$\text{Re } V(\mathbf{r}, t) = -\frac{1}{\pi} P \int_{-\infty}^{\infty} \frac{\text{Im } V(\mathbf{r}, t')}{t' - t} dt',$$

where P denotes Cauchy's principal value.

The use of the analytic signal V instead of $V^{(r)}$ has the advantage that it is closely connected with the quantum-mechanical description, as will be seen later. Furthermore, it is convenient to define the *instantaneous intensity* by

$$I(\mathbf{r}, t) = V^*(\mathbf{r}, t) V(\mathbf{r}, t).$$

(1.3.7)

This does not represent the instantaneous intensity of the real field $V^{(r)}$. In fact, for a monochromatic or quasi-monochromatic field, $[V^{(r)}]^2$ is a rapidly oscillating function of time, with period of the order of the mean optical periods, i.e., 10^{-15} sec. This is several orders of magnitude shorter than the resolving time of the fastest optical detectors available, so that the actual intensity registered by the detector corresponds to an average over many periods of oscillation, and this may readily be shown to be represented by (1.3.7).

In view of the unavoidable fluctuations mentioned in § 1.2, a real light field is not characterized by a given function $V(\mathbf{r}, t)$, but rather by a stochastic variable, i.e., a member of a statistical ensemble. Each member corresponds to a possible realization of the field. If $p_1(V)$ is the probability distribution associated with $V(\mathbf{r}, t)$ in the ensemble, i.e., if the probability of finding for the field at $\mathbf{r}$, t a value between $V = \text{Re } V + i \text{ Im } V$ and $\text{Re } V + d(\text{Re } V) + i \text{ Im } V + i d(\text{Im } V)$ is $p_1(V)d^2 V = p_1(V)d(\text{Re } V) d(\text{Im } V)$, the ensemble average $\langle f(V) \rangle$ of any function of V is given by

$$\langle f(V) \rangle = \int f(V) p_1(V) d^2 V,$$

(1.3.8)

where the integral is extended over the whole complex V-plane.

In particular, the *average intensity* at $(\mathbf{r}, t)$ is given by

$$\langle I(\mathbf{r}, t) \rangle = \langle V^*(\mathbf{r}, t) V(\mathbf{r}, t) \rangle.$$

(1.3.9)

(b)   Second-order coherence functions

Let us apply the above results to the evaluation of the average intensity distribution in Young's interference experiment (Fig. 1.2). The wave function at $\mathbf{r}$ at the time t (for a given realization of the field) is a linear function of its corresponding retarded values at the pinholes:

$$V(\mathbf{r}, t) = K_1 V(\mathbf{r}_1, t - t_1) + K_2 V(\mathbf{r}_2, t - t_2), \qquad (1.3.10)$$

where $t_1 = d_1/c$, $t_2 = d_2/c$, and the geometrical factors K_1 and K_2 are purely imaginary, according to classical diffraction theory[11].

It follows that

$$I(\mathbf{r}, t) = V^*(\mathbf{r}, t)\, V(\mathbf{r}, t) = |K_1|^2\, I(\mathbf{r}_1, t - t_1) + |K_2|^2\, I(\mathbf{r}_2, t - t_2)$$
$$+ 2K_1^* K_2\, \text{Re}\, [V^*(\mathbf{r}_1, t - t_1)\, V(\mathbf{r}_2, t - t_2)], \qquad (1.3.11)$$

where the last term represents the interference effects. Taking the ensemble average, we get

$$\langle I(\mathbf{r}, t) \rangle = |K_1|^2 \langle I(\mathbf{r}_1, t - t_1) \rangle + |K_2|^2 \langle I(\mathbf{r}_2, t - t_2) \rangle$$
$$+ 2|K_1 K_2|\, \text{Re}\, \Gamma(\mathbf{r}_1, t - t_1; \mathbf{r}_2, t - t_2), \qquad (1.3.12)$$

where

$$\Gamma(\mathbf{r}_1, t_1; \mathbf{r}_2, t_2) = \langle V^*(\mathbf{r}_1, t_1)\, V(\mathbf{r}_2, t_2) \rangle. \qquad (1.3.13)$$

This function, which expresses the cross-correlation between the fields at two different space-time points, is called the *mutual coherence function*. It is one of the basic quantities in the classical theory of coherence.

Usually one deals with *stationary fields*. This means that the probability distribution $p_1(V)$ in (1.3.8) is invariant under time translations, so that all ensemble averages are independent of the choice of the origin of time. In particular, the average intensity (1.3.9) is time-independent. Furthermore, (1.3.13) depends only on the difference $t_2 - t_1 = \tau$:

$$\Gamma(\mathbf{r}_1, t; \mathbf{r}_2, t + \tau) = \Gamma(\mathbf{r}_1, \mathbf{r}_2, \tau) = \langle V^*(\mathbf{r}_1, t)\, V(\mathbf{r}_2, t + \tau) \rangle. \qquad (1.3.14)$$

Note that $\Gamma(\mathbf{r}, \mathbf{r}, 0) = \langle I(\mathbf{r}, t) \rangle$ [cf. (1.3.9)]. Assuming that the field is also *ergodic*, one may replace ensemble averages by time averages,

$$\langle f(\mathbf{r}, t) \rangle \rightarrow \lim_{T \to \infty} \frac{1}{2T} \int_{-T}^{T} f(\mathbf{r}, t)\, dt. \qquad (1.3.15)$$

For a stationary field, the result (1.3.12) becomes

$$\langle I(\mathbf{r}) \rangle = |K_1|^2 \langle I(\mathbf{r}_1) \rangle + |K_2|^2 \langle I(\mathbf{r}_2) \rangle$$
$$+ 2|K_1 K_2|\, \text{Re}\, \Gamma(\mathbf{r}_1, \mathbf{r}_2, \tau), \qquad (1.3.16)$$

where $\tau = (d_2 - d_1)/c$, and the average intensities are now time-independent.

It is convenient to introduce the normalized quantity

$$\gamma(\mathbf{r}_1, \mathbf{r}_2, \tau) = \frac{\Gamma(\mathbf{r}_1, \mathbf{r}_2, \tau)}{\sqrt{\langle I(\mathbf{r}_1)\rangle \langle I(\mathbf{r}_2)\rangle}}, \tag{1.3.17}$$

known as the *complex degree of coherence*. It follows from (1.3.9), (1.3.14), (1.3.15) and Schwarz's inequality that

$$0 \leq |\gamma(\mathbf{r}_1, \mathbf{r}_2, \tau)| \leq 1. \tag{1.3.18}$$

The result (1.3.16) then becomes

$$\langle I(\mathbf{r})\rangle = \langle I_1\rangle + \langle I_2\rangle + 2\sqrt{\langle I_1\rangle \langle I_2\rangle}\,\mathrm{Re}\,\gamma(\mathbf{r}_1, \mathbf{r}_2, \tau), \tag{1.3.19}$$

where

$$\langle I_j\rangle = |K_j|^2 \langle I(\mathbf{r}_j)\rangle \quad (j = 1, 2) \tag{1.3.20}$$

is the average intensity that one would observe at $\mathbf{r}$ if only the hole j were open.

Let

$$\gamma(\mathbf{r}_1, \mathbf{r}_2, \tau) = |\gamma(\mathbf{r}_1, \mathbf{r}_2, \tau)| \exp\,[i\varphi(\mathbf{r}_1, \mathbf{r}_2, \tau)], \tag{1.3.21}$$

and let us assume, for simplicity, that the pinholes are identical and that the geometry is such that

$$\langle I_1\rangle \approx \langle I_2\rangle \approx I. \tag{1.3.22}$$

The interference law (1.3.19) becomes

$$\langle I(\mathbf{r})\rangle = 2I[1 + |\gamma(\mathbf{r}_1, \mathbf{r}_2, \tau)|\cos\varphi(\mathbf{r}_1, \mathbf{r}_2, \tau)]. \tag{1.3.23}$$

In the ideal case of strictly monochromatic light, $V(\mathbf{r}) = |V(\mathbf{r})|\,e^{-2\pi i\nu t}$, it follows from (1.3.14) and (1.3.17) that

$$\gamma(\mathbf{r}_1, \mathbf{r}_2, \tau) = e^{-2\pi i\nu\tau}, \quad |\gamma| = 1, \tag{1.3.24}$$

and (1.3.23) leads to the well-known Young interference fringes of the elementary theory (Fig. 1.3). The light is then said to be *coherent*. In the opposite extreme case, $\gamma = 0$, we see from (1.3.23) or (1.3.19) that there is no interference (the intensities are added); we then have *incoherent* light. In the intermediate case $0 < |\gamma| < 1$, we speak of *partially coherent* light.

For *quasi-monochromatic light*, i.e., light whose spectral components have a bandwidth $\Delta\nu \ll \bar{\nu}$, where $\bar{\nu}$ is the central frequency, it is readily seen that $|\gamma(\mathbf{r}_1, \mathbf{r}_2, \tau)|$ plays the role of a slowly-varying envelope function, whereas $\varphi(\mathbf{r}_1, \mathbf{r}_2, \tau) \approx -2\pi\bar{\nu}\tau$ for times that are short as compared with the coherence time $1/\Delta\nu$. Thus we still obtain interference fringes as for (1.3.24) (Fig. 1.3), but with intensity maxima and minima given by

$$\langle I\rangle_{\max} = 2I(1 + |\gamma|); \quad \langle I\rangle_{\min} = 2I(1 - |\gamma|). \tag{1.3.25}$$

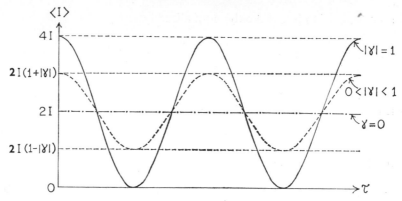

Figure 1.3 Interference fringes and degree of coherence for coherent light ($|\gamma| = 1$), partially coherent light ($0 < |\gamma| < 1$), and incoherent light ($\gamma = 0$)

The contrast of the interference fringes may be measured by defining their *visibility* $\mathscr{V}$ as

$$\mathscr{V} = \frac{\langle I \rangle_{\text{max}} - \langle I \rangle_{\text{min}}}{\langle I \rangle_{\text{max}} + \langle I \rangle_{\text{min}}}. \qquad (1.3.26)$$

It then follows from (1.3.25) that

$$\mathscr{V} = |\gamma|, \qquad (1.3.27)$$

i.e., the visibility of the interference fringes is a direct measure of the magnitude of the degree of coherence.

According to (1.3.23), the positions of the maxima of the interference fringes are determined by

$$\varphi(\mathbf{r}_1, \mathbf{r}_2, \tau) = 2m\pi \quad (m = 0, \pm 1, \pm 2, ...). \qquad (1.3.28)$$

This indicates how one can measure, in principle, the phase of the degree of coherence, but this is usually a much more rapidly varying quantity than $|\gamma|$, so that it is considerably more difficult to measure.

(c) Interference spectroscopy and stellar interferometry

The interference fringes in Michelson's interferometer (Fig. 1.1) can be described by (1.3.23), with $\mathbf{r}_1 = \mathbf{r}_2$ corresponding to the point 0 at which the beam is split. The observed fringes are determined by Re $\gamma(\tau)$, where

$$\gamma(\tau) = \gamma(\mathbf{r}, \mathbf{r}, \tau) = \frac{\Gamma(\mathbf{r}, \mathbf{r}, \tau)}{\Gamma(\mathbf{r}, \mathbf{r}, 0)} \qquad (1.3.29)$$

is called the *complex degree of temporal coherence* (the dependence on $\mathbf{r}$ is frequently omitted).

By (1.3.14), we have (omitting the spatial dependence)

$$\Gamma(\tau) = \langle V^*(t) \, V(t + \tau) \rangle. \tag{1.3.30}$$

Inserting the Fourier representation (1.3.3), we find

$$\Gamma(\tau) = \int_0^\infty dv \int_0^\infty dv' \langle v^*(v) \, v(v') \rangle \, e^{-2\pi i v' \tau} e^{-2\pi i (v'-v)t}. \tag{1.3.31}$$

For a *stationary field*[12], the right-hand side must be independent of t, which is only possible if

$$\langle v^*(v) \, v(v') \rangle = G(v) \, \delta(v - v'), \tag{1.3.32}$$

where δ is the Dirac delta function. Thus,

$$\Gamma(\tau) = \int_0^\infty G(v) \, e^{-2\pi i v \tau} \, dv. \tag{1.3.33}$$

In particular, for $\tau = 0$, $\Gamma(\tau)$ represents the average intensity, as we see from (1.3.9). Thus,

$$\langle I \rangle = \Gamma(0) = \int_0^\infty G(v) \, dv, \tag{1.3.34}$$

showing that $G(v)$ represents the *spectral density* (also known as the *power spectrum* in the theory of random processes). The result (1.3.33), which corresponds to the well-known *Wiener-Khintchine theorem* in the theory of stationary random processes, asserts that the *spectral density and the complex degree of temporal coherence are Fourier transforms of each other.* This allows one to give a more precise justification for the relation (1.2.1) between spectral width and coherence time, by the same arguments employed in connection with the uncertainty relation. For example, if the spectrum is Lorentzian,

$$G(v) = \frac{A}{(v - v_0)^2 + (\Delta v)^2}, \tag{1.3.35}$$

we find, for times τ not too large in comparison with $1/\Delta v$,

$$\gamma(\tau) \approx e^{-2\pi i v_0 \tau - \Delta v \cdot \tau}, \tag{1.3.36}$$

so that the degree of coherence decays exponentially, with lifetime given by the coherence time $1/\Delta v$.

The relation (1.3.33) is the basis of Michelson's method of *interference spectroscopy*[13]. If one measures $\operatorname{Re} \Gamma(\tau)$ with Michelson's interferometer, one can obtain the spectrum, according to (1.3.33), by taking the Fourier cosine transform. In practice, for quasi-monochromatic light in the optical region, it is much easier to measure $|\gamma|$ (visibility) than the phase φ, as has already been mentioned. Since both are needed to reconstruct the spectrum, the problem arises of trying to determine the phase, given the modulus. These two quantities are not independent: by (1.3.33), $\Gamma(\tau)$ is also an analytic signal, so that modulus and phase are connected by dispersion relations analogous to (1.3.6). Unfortunately, although one can make use of this analyticity to solve the phase problem in some special cases[14], this does not seem to be possible in general[15].

It follows immediately from the definition (1.3.13) and from the wave equation for the field $V(\mathbf{r}, t)$ that

$$\square_j \Gamma(\mathbf{r}_1, t_1; \mathbf{r}_2, t_2) = \left(\varDelta_j - \frac{1}{c^2} \frac{\partial^2}{\partial t_j^2} \right) \Gamma(\mathbf{r}_1, t_1; \mathbf{r}_2, t_2) = 0 \quad (j = 1, 2),$$

(1.3.37)

i.e., Γ obeys two wave equations, one in each argument. Light which is initially incoherent at the source may acquire spatial coherence by the process of propagation: this explains why light from a distant star can form diffraction fringes in the focal plane of a telescope.

In particular, for a stationary field, with

$$\Gamma(\mathbf{r}, \mathbf{r}', \tau) = \int\limits_0^\infty G(\mathbf{r}, \mathbf{r}', \nu) \, e^{-2\pi i \nu \tau} \, d\nu,$$

(1.3.38)

the monochromatic components $G(\mathbf{r}, \mathbf{r}', \nu)$ satisfy the Helmholtz wave equation in each argument. A typical boundary-value problem would be that of determining G given its values when $\mathbf{r}$ and $\mathbf{r}'$ are both points of a given surface S (e.g., the surface of the source). An ideal source with complete spatial incoherence, for instance, would be such that no correlation exists between emissions from any two different source points ϱ and ϱ', so that

$$G(\varrho, \varrho', \nu) = \delta(\varrho - \varrho') \, G(\varrho, \varrho, \nu) = \delta(\varrho - \varrho') \, \langle I(\varrho, \nu) \rangle \quad (\varrho \in S, \varrho' \in S),$$

(1.3.39)

where $\langle I(\varrho, \nu) \rangle$ is the Fourier component of frequency ν of the average intensity at the point ϱ of the source [cf. (1.3.34)].

The associated boundary-value problem is entirely similar to that for the wave function, and it can be solved by well-known Green's function techniques[16]. Accordingly, the result can be physically interpreted in terms of Huygens' principle, and it is equivalent to a diffraction pattern produced by an aperture of the same size and shape as the source. The precise formulation of this result is contained in the *van Cittert-Zernike theorem*[17].

In particular, if the distance between the source and the region of observation is large, we get the analogue of a Fraunhofer diffraction pattern, and, as is well-known[18], this corresponds to the Fourier transform of the spatial distribution over the aperture. By (1.3.39), this leads to

$$G(\mathbf{r}, \mathbf{r}', \nu) = C \int_S \exp\left[-ik\boldsymbol{\varrho} \cdot (\hat{\mathbf{u}} - \hat{\mathbf{u}}')\right] \langle I(\boldsymbol{\varrho}, \nu) \rangle \, d^2\varrho, \qquad (1.3.40)$$

where $\hat{\mathbf{u}}$ and $\hat{\mathbf{u}}'$ are unit vectors in the directions of $\mathbf{r} - \boldsymbol{\varrho}$ and $\mathbf{r}' - \boldsymbol{\varrho}$, respectively. For quasi-monochromatic light, this result holds true also for the mutual coherence function. Thus *the mutual coherence function at large distances from an incoherent source is proportional to the Fourier transform of the intensity distribution over the source.*

If the source can be assimilated to a circular disc of radius a with uniform intensity distribution, we get the well-known Airy diffraction pattern of a circular disc: a bright central disc, surrounded by faint concentric rings. Most of the intensity is concentrated in the central disc, whose angular radius is given by the well-known formula

$$\sin \theta \approx 0.61\lambda/a, \qquad (1.3.41)$$

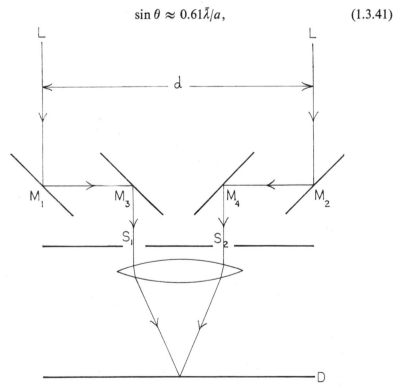

Figure 1.4 Michelson's stellar interferometer. L: starlight; M_1, M_2: movable mirrors; M_3, M_4: fixed mirrors; S_1, S_2: slits; D: detector

where $\bar{\lambda}$ is the average wavelength. If R is the distance from the source to the observation plane, the radius d of the coherence area (§ 1.2) on this plane may therefore be taken as

$$d \sim R \sin \theta \approx 0.61 \bar{\lambda}/\alpha, \qquad (1.3.42)$$

where $\alpha = a/R$ is the apparent angular radius of the source viewed from the observation plane.

In Michelson's stellar interferometer (Fig. 1.4), these results were employed to measure the angular diameters of stars. Light from the star falls on the movable mirrors, M_1, M_2 [corresponding to the points $\mathbf{r}$, $\mathbf{r}'$ in (1.3.40)] and is directed by the fixed mirrors M_3, M_4 to a detector D in the focal plane of a telescope. The separation d between the movable mirrors is increased until the visibility of the interference fringes goes to zero for the first time; this value corresponds to (1.3.42) and allows α to be determined. In this way Michelson was able to measure angular diameters down to 0.02''. However, this required a base line d of several meters, and problems of mechanical stability as well as random fluctuations in the optical path due to atmospheric effects make it very difficult to go beyond this limit.

(d)   Higher-order coherence functions

In a series of experiments beginning in 1955, Hanbury Brown and Twiss[19] demonstrated the possibility of measuring the magnitude of the coherence function by means of the correlation between *intensity fluctuations* at two different points, rather than the correlation between the fields. Their *stellar intensity interferometer* is schematically shown in Fig. 1.5. Light from a star is collected by two parabolic reflectors and focussed on two phototubes P_1 and P_2. The amplified output currents are fed into an electronic "correlator", which allows one to measure

$$\langle \Delta I(\mathbf{r}_1, t)\, \Delta I(\mathbf{r}_2, t + \tau)\rangle, \qquad (1.3.43)$$

i.e., the correlation between *intensity fluctuations* at P_1 and P_2.

As will be seen later [cf. § 4.3(c)], this quantity is proportional to $|\Gamma(\mathbf{r}_1, \mathbf{r}_2, \tau)|^2$ so that, as in Michelson's stellar interferometer, it can be used to determine angular diameters of stars. However, since the intensity fluctuations are slowly-varying as compared with the optical frequencies, this method is far less sensitive to atmospheric and mechanical disturbances. Thus, in the correlation interferometer recently built at Narrabri (Australia), the base line reaches about 200 meters, allowing the measurement of angular diameters down to 0.0005''!

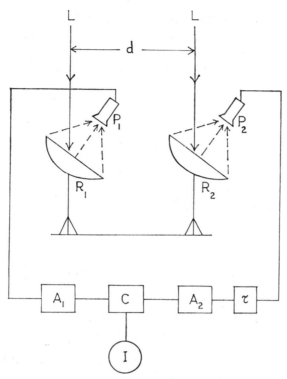

Figure 1.5 Hanbury Brown and Twiss stellar interferometer. L: starlight; R_1, R_2: parabolic reflectors; P_1, P_2: phototubes; τ: delay line; A_1, A_2: amplifiers; C: correlator; I: integrator

The Hanbury Brown and Twiss experiment was the first observation of *higher-order coherence*. The correlation function (1.3.43) is of the type

$$\langle I_1 I_2 \rangle = \langle V^*(\mathbf{r}_1, t_1) \, V(\mathbf{r}_1, t_1) \, V^*(\mathbf{r}_2, t_2) \, V(\mathbf{r}_2, t_2) \rangle. \qquad (1.3.44)$$

More generally, one can define higher-order coherence functions of the type

$$\Gamma^{(m,n)}(x_1, x_2, \ldots, x_m; x_{m+1}, \ldots, x_{m+n})$$

$$= \langle V^*(x_1) \ldots V^*(x_m) \, V(x_{m+1}) \ldots V(x_{m+n}) \rangle, \qquad (1.3.45)$$

where

$$x_j = (\mathbf{r}_j, t_j) \qquad (j = 1, 2, \ldots, m+n). \qquad (1.3.46)$$

This is called a *coherence function of order* $N = m + n$. Thus the functions discussed in sections (b) and (c) were *second-order* coherence functions.

The ensemble average in (1.3.45) is defined by [cf. (1.3.8)]

$$\langle f(V_1, \ldots, V_N) \rangle = \int f(V_1, \ldots, V_N) \, p_N(V_1, \ldots, V_N) \, d^2 V_1 \ldots d^2 V_N, \qquad (1.3.47)$$

where $p_N(V_1, ..., V_N)$ is the *joint probability distribution* of the random variables $V_1, ..., V_N$, and the integral is extended over N complex planes.

In particular, if the field is stationary, p_N is invariant under time translations, and (1.3.45) depends only on $N - 1$ time variables, e.g., the differences $\tau_j = t_j - t_1$ $(j = 2, 3, ..., N)$.

When polarization effects are taken into account, V has to be replaced by a vector quantity (e.g., vector potential or electric field), so that the coherence functions (1.3.45) become higher-rank tensors containing a vector (polarization) index for each argument. For the treatment of partially polarized light, as well as for a more detailed treatment of classical coherence theory, the reader is referred to Born and Wolf's book[8] and to the review article by Mandel and Wolf[20].

References

1. *Quantum Electronics and Coherent Light* (hereafter referred to as *QECL*), *Proceedings of the International School of Physics "Enrico Fermi"*, Course 31, edited by P. A. Miles, Academic Press, New York (1964).
2. *Quantum Optics and Electronics* (hereafter referred to as *QOE*), *Les Houches 1964*, edited by C. De Witt, A. Blandin, and C. Cohen-Tannoudji, Gordon and Breach, New York (1965).
3. *Quantum Optics* (hereafter referred to as *QO*), *Proceedings of the International School of Physics "Enrico Fermi"*, *Course 42*, edited by R. J. Glauber, Academic Press, New York (1970).
4. J. R. Klauder and E. C. G. Sudarshan, *Fundamentals of Quantum Optics* (hereafter referred to as *KS*), W. A. Benjamin, New York (1968).
5. For an extensive bibliography and collection of reprints, cf. *Selected Papers on Coherence and Fluctuations of Light*, edited by L. Mandel and E. Wolf, vols. I and II, Dover Publications, New York (1970).
6. R. Hanbury Brown and R. Q. Twiss, *Proc. Roy. Soc. (London)*, A**242**, 300 (1957); L. Mandel and E. Wolf, *Rev. Mod. Phys.* **37**, 231 (1965).
7. L. Mandel, *J. Opt. Soc. Am.* **51**, 797 (1961).
8. For a more complete treatment of the classical theory of coherence, see M. Born and E. Wolf, *Principles of Optics*, Pergamon Press, London (1959), Chapter X.
9. D. Gabor, *J. Inst. Electr. Engrs.* **93**, 429 (1946).
10. Cf., e.g., H. M. Nussenzveig, *Causality and Dispersion Relations*, Academic Press, New York (1972).
11. M. Born and E. Wolf, loc. cit., § 8.3.
12. For a stationary field, the Fourier decomposition (1.3.3) does not exist as the integral of an ordinary function. The results have to be interpreted in the sense of generalized random functions; cf. I. M. Gel'fand and N. Ya. Vilenkin, *Generalized Functions*, Vol. 4, Academic Press, New York (1964), Chapter III. Alternatively, one can introduce truncated functions of time, vanishing for $|t| > T$, and at the end go over to the limit $T \to \infty$, as was done in Wiener's pioneering work on generalized harmonic analysis [*Acta Math.* **55**, 117 (1930)].
13. M. Born and E. Wolf, loc. cit., §§ 7.5.8 and 10.4.1.

14. E. Wolf, *Proc. Phys. Soc.* (*London*) **80**, 1269 (1962); Y. Kano and E. Wolf, *ibid.*, 1273 (1962).
15. H. M. Nussenzveig, *J. Math. Phys.* **8**, 561 (1967).
16. M. Born and E. Wolf, ibid., § 10.4.2.
17. P. H. van Cittert, *Physica* **1**, 201 (1934); F. Zernike, *Physica* **5**, 785 (1938).
18. C. J. Bouwkamp, *Rep. Progr. Phys.* **17**, 35 (1954).
19. R. Hanbury Brown and R. Q. Twiss, *Nature* **177**, 27 (1956); **178**, 1046, 1447 (1956).
20. L. Mandel and E. Wolf, *Rev. Mod. Phys.* **37**, 231 (1965).

CHAPTER 2

Quantum Coherence Functions

2.1 THE QUANTIZED RADIATION FIELD

WE ASSUME THAT the reader is familiar with the quantization of the free electromagnetic radiation field[1]. In the *radiation gauge*, the scalar potential $\varphi = 0$ and the vector potential $\mathbf{A}$ satisfies the transversality condition,

$$\nabla \cdot \mathbf{A} = 0. \tag{2.1.1}$$

If we expand the field within a cubical box of side L with periodic boundary conditions, the quantized radiation field is described by the vector potential operator $\mathbf{A}(\mathbf{r}, t)$ with

$$\mathbf{A}(\mathbf{r}, t) = \frac{1}{L^{3/2}} \sum_k \frac{1}{\sqrt{2\omega_k}} [a_k \varepsilon_k e^{i(\mathbf{k}\cdot\mathbf{r} - \omega_k t)} + a_k^+ \varepsilon_k^* e^{-i(\mathbf{k}\cdot\mathbf{r} - \omega_k t)}]$$

$$= \mathbf{A}^{(+)}(\mathbf{r}, t) + \mathbf{A}^{(-)}(\mathbf{r}, t), \tag{2.1.2}$$

where

$$\mathbf{k} = \frac{2\pi}{L}(n_1, n_2, n_3), \quad \text{with} \quad n_i = 0, \pm 1, \pm 2, \ldots;$$

$\omega_k = |\mathbf{k}|$ (in units $\hbar = c = 1$), and the index k stands for

$$k \to (\mathbf{k}, s), \tag{2.1.3}$$

with the polarization index s taking the values $s = 1, 2$.

The polarization vectors $\varepsilon_k = \varepsilon_{k,s}$ are two transverse unit vectors,

$$\mathbf{k} \cdot \varepsilon_{\mathbf{k},s} = 0, \tag{2.1.4}$$

$$\varepsilon_{\mathbf{k},s}^* \cdot \varepsilon_{\mathbf{k},s'} = \delta_{s,s'}. \tag{2.1.5}$$

They may be two real vectors $\varepsilon_1, \varepsilon_2$, but one may also take them as complex, e.g.,

$$\varepsilon_{(\pm)} = \frac{1}{\sqrt{2}}(\varepsilon_1 \pm i\varepsilon_2), \tag{2.1.6}$$

which describe orthogonal circular polarizations[2].

The operators a_k and a_k^+ are *annihilation and creation operators*, respectively, for photons with momentum **k** and polarization s. They satisfy the commutation rules

$$[a_k, a_{k'}] = [a_k^+, a_{k'}^+] = 0, \tag{2.1.7}$$

$$[a_k, a_{k'}^+] = \delta_{k,k'} = \delta_{\mathbf{k,k'}} \delta_{s,s'}. \tag{2.1.8}$$

The annihilation part $\mathbf{A}^{(+)}(\mathbf{r}, t)$ contains only positive frequencies ($e^{-i\omega_k t}$, $\omega_k > 0$); its hermitian conjugate, $\mathbf{A}^{(-)}(\mathbf{r}, t) = [\mathbf{A}^{(+)}(\mathbf{r}, t)]^+$, contains only negative frequencies ($e^{i\omega_k t}$, $\omega_k > 0$). As will be seen later, the positive-frequency part plays a role analogous to that of the analytic signal V in the classical theory [cf. (1.3.3)].

The quantized electric field $\mathbf{E}(\mathbf{r}, t)$ is given by

$$\mathbf{E}(\mathbf{r}, t) = -\dot{\mathbf{A}}(\mathbf{r}, t) = \frac{i}{L^{3/2}} \sum_k \sqrt{\frac{\omega_k}{2}} [a_k \varepsilon_k e^{i(\mathbf{k\cdot r} - \omega_k t)} - a_k^+ \varepsilon_k^* e^{-i(\mathbf{k\cdot r} - \omega_k t)}]$$

$$= \mathbf{E}^{(+)}(\mathbf{r}, t) + \mathbf{E}^{(-)}(\mathbf{r}, t). \tag{2.1.9}$$

If we remove the restriction to a box, (2.1.2) goes over into a plane-wave expansion, with **k** varying continuously; we then have to make the substitutions

$$\frac{1}{L^{3/2}} \sum_k = \frac{1}{L^{3/2}} \sum_{k,s} \to \frac{1}{(2\pi)^{3/2}} \sum_s \int d^3k, \tag{2.1.10}$$

$$\delta_{k,k'} \to \delta(\mathbf{k} - \mathbf{k'}) \delta_{s,s'}. \tag{2.1.11}$$

A *pure state* of the radiation field is described by a normalized state vector $|\psi\rangle$,

$$\langle \psi \mid \psi \rangle = 1, \tag{2.1.12}$$

where $|\psi\rangle$ may be thought of, for instance, as a superposition of states with definite numbers of photons (Fock states). The expectation value of an operator $\mathcal{O}$ associated with some physical quantity in this state is

$$\langle \mathcal{O} \rangle = \langle \psi | \mathcal{O} | \psi \rangle. \tag{2.1.13}$$

As a rule, of course, the value of the corresponding physical quantity still undergoes quantum fluctuations about this average, unless $|\psi\rangle$ is an eigenstate of $\mathcal{O}$.

In general, our knowledge of the field is not sufficiently complete for us to describe it by a pure state. As in the classical case (cf. § 1.3), all we can say is that it is a member of a statistical ensemble of (pure) states $|\psi_n\rangle$, with a probability (statistical weight) p_n to be found in the state $|\psi_n\rangle$. We then have a *mixed state*, and the expectation value of an operator $\mathcal{O}$ in this state is obtained by combining the quantum average (2.1.13) with a statistical average over the weights p_n:

$$\langle \mathcal{O} \rangle = \sum_n p_n \langle \psi_n | \mathcal{O} | \psi_n \rangle. \tag{2.1.14}$$

The field can then be described by the *density operator*[3]

$$\varrho = \sum_n p_n | \psi_n \rangle \langle \psi_n |, \tag{2.1.15}$$

where $| \psi_n \rangle \langle \psi_n |$ is the projector over the state $| \psi_n \rangle$. Note that

$$\varrho_{mn} = \langle \psi_m | \varrho | \psi_n \rangle = p_n \delta_{mn}.$$

In terms of ϱ, the expectation value (2.1.14) may be rewritten as

$$\langle \mathcal{O} \rangle = \mathrm{Tr}\,(\varrho \mathcal{O}) = \sum_{m,n} \varrho_{mn} \mathcal{O}_{nm}, \tag{2.1.16}$$

where the trace (Tr) of an operator is the sum of its diagonal matrix elements in any representation; it has the property

$$\mathrm{Tr}\,(AB) = \mathrm{Tr}\,(BA). \tag{2.1.17}$$

The expression (2.1.16) is the analogue of the classical statistical average

$$\overline{\mathcal{O}} = \int \varrho \mathcal{O} \, d\mu, \tag{2.1.18}$$

where ϱ is the density function representing a classical ensemble.

As the weights p_n represent probabilities, we have

$$0 \leq p_n \leq 1, \tag{2.1.19}$$

$$\mathrm{Tr}\,\varrho = \sum_n p_n = 1. \tag{2.1.20}$$

Furthermore, by (2.1.15), ϱ is hermitian,

$$\varrho^+ = \varrho. \tag{2.1.21}$$

A particular case of (2.1.15) is the density operator representing a pure state $| \psi \rangle$,

$$\varrho = | \psi \rangle \langle \psi |, \tag{2.1.22}$$

which is characterized by the property

$$\varrho^2 = | \psi \rangle \langle \psi | \psi \rangle \langle \psi | = | \psi \rangle \langle \psi | = \varrho. \tag{2.1.23}$$

If we change the basis in (2.1.15),

$$| \psi_n \rangle = \sum_m c_{nm} | \varphi_m \rangle, \tag{2.1.24}$$

2*

the density operator becomes

$$\varrho = \sum_{i,j} \varrho_{ij} |\varphi_i\rangle \langle\varphi_j|, \tag{2.1.25}$$

where

$$\varrho_{ij} = \langle\varphi_i| \varrho |\varphi_j\rangle = \sum_n p_n c_{ni} c_{nj}^*. \tag{2.1.26}$$

The representation (2.1.15), which is simpler than the general representation (2.1.25) of the density operator (it involves a single sum instead of a double sum), is called a *diagonal representation*. Note that

$$\varrho_{ii} \geqq 0, \tag{2.1.27}$$

i.e., the diagonal elements are non-negative in *any* representation; for this reason, ϱ is called a *positive* operator.

2.2   THEORY OF PHOTOELECTRIC DETECTION

(a)   Introduction

In the quantum theory of coherence, as originally formulated by Glauber[4], the analysis of the process of detection of light plays a basic role. Most detection processes (including the visual and photographic processes) are based on the photoelectric effect, and the quantized character of the field is manifested through the quantum nature of the photoelectric effect.

The elementary process involved in the detection is the absorption of a photon, with the corresponding emission of a photoelectron. In principle, if a detector atom is initially in an excited state, we might also have a sort of converse process, in which induced emission of a photon takes place under the action of the incident field, and the atom returns to the ground state. However, for most detectors in usual circumstances, nearly all atoms are in the ground state, so that only absorption can take place [cf., however, § 2.2(d)].

Since it is only the annihilation part $E^{(+)}$ of the field (2.1.9) that plays a significant role in absorption, the detection process introduces a fundamental asymmetry between $E^{(+)}$ and $E^{(-)}$, in such a way that what one actually detects corresponds more closely to $E^{(+)}$ than to the real field E. This would no longer be true in the classical limit: if $\hbar\omega \to 0$, a test body placed in the field would emit photons as well as absorb them; $E^{(+)}$ and $E^{(-)}$ would play comparable roles, and one would measure the real field.

An ideal photodetector would have negligible spatial extension and wide-band sensitivity, thus enabling it to respond to the field at a single point $\mathbf{r}$ in space at a definite instant of time t. The transition amplitude for a process in which a photon is absorbed at $(\mathbf{r}, t)$ from the field in the initial state $|\psi_i\rangle$, leading to the final state $|\psi_f\rangle$, should be proportional to the matrix element

$$\langle\psi_f|\,E^{(+)}(\mathbf{r},\,t)\,|\psi_i\rangle,$$

where $E^{(+)}$ is the annihilation part of the electric field, defined by (2.1.9), and we assume for simplicity that there is only one polarization component. As will be seen below, this is actually an approximation, based on the use of first-order perturbation theory and on the dipole approximation, but it is a very good approximation.

In general one does not determine the final state of the field: all possible final states contribute to the total counting rate. Thus, the counting rate is obtained by summing the transition probability per unit time over all accessible final states $|\psi_f\rangle$, differing from $|\psi_i\rangle$ by the absorption of a photon,

$$
\begin{aligned}
w_i &= \sum_f w_{i\rightarrow f} = \sum_f |\langle\psi_f|\,E^{(+)}(\mathbf{r},\,t)\,|\psi_i\rangle|^2 \\
&= \sum_f \langle\psi_i|\,E^{(-)}(\mathbf{r},\,t)\,|\psi_f\rangle\,\langle\psi_f|\,E^{(+)}(\mathbf{r},\,t)\,|\psi_i\rangle. \qquad (2.2.1)
\end{aligned}
$$

We may as well, however, sum over a complete set of final states, because final states differing from $|\psi_i\rangle$ by more than the absorption of one photon must be orthogonal to $E^{(+)}|\psi_i\rangle$ and, therefore, do not contribute to the sum. With the help of the completeness relation

$$\sum_f |\psi_f\rangle\,\langle\psi_f| = 1, \qquad (2.2.2)$$

the result (2.2.1) becomes

$$w_i = \langle\psi_i|\,E^{(-)}(\mathbf{r},\,t)\,E^{(+)}(\mathbf{r},\,t)\,|\psi_i\rangle. \qquad (2.2.3)$$

The initial state of the field is not usually a pure state $|\psi_i\rangle$, but rather a mixed state, described by a density operator

$$\varrho = \sum_i p_i\,|\psi_i\rangle\,\langle\psi_i|. \qquad (2.2.4)$$

In this case we have to replace (2.2.3) by the statistical average,

$$\langle w_i\rangle = \sum_i p_i w_i = \mathrm{Tr}\,[\varrho E^{(-)}(\mathbf{r},\,t)\,E^{(+)}(\mathbf{r},\,t)]. \qquad (2.2.5)$$

The counting rate of an ideal photodetector is therefore proportional to the quantity (2.2.5), which may be taken as a measure of the *average intensity* at $(\mathbf{r},\,t)$. By comparison with (1.3.9), we infer that the quantum analogue of the second-order coherence function (1.3.13) should be

$$G^{(1,1)}(x,\,x') = \mathrm{Tr}\,[\varrho E^{(-)}(x)\,E^{(+)}(x')], \qquad (2.2.6)$$

where [cf. (1.3.46)] $x = (\mathbf{r},\,t)$; $x' = (\mathbf{r}',\,t')$. In particular, the average intensity is measured by $G^{(1,1)}(x,\,x)$.

(b) Single-atom detector

Let us now present a more detailed version of the argument given in the last section, about the detection of a photon by an ideal photodetector[5]. We assume for simplicity that the detector is represented by a single atom with only one valence electron, that interacts with the radiation field by making a transition from the bound (ground) state $|g\rangle$ to a discrete excited state $|e\rangle$.

The hamiltonian of the system may be written as

$$H = H_0 + H_I = H_A + H_F + H_I, \qquad (2.2.7)$$

where the unperturbed hamiltonian H_0 is the sum of the atomic hamiltonian H_A and the hamiltonian of the free radiation field H_F.

We use the interaction picture, in which operators are time-dependent, with their time dependence given by the unperturbed hamiltonian:

$$H_I(t) = e^{iH_0 t} H_I^{(S)} e^{-iH_0 t}, \qquad (2.2.8)$$

where $H_I^{(S)}$ is the interaction hamiltonian in the Schrödinger picture. Let the electric field be linearly polarized in the x-direction, and let the origin of coordinates be taken at some fixed point in the atom. Since optical wavelengths are much greater than atomic dimensions, we may employ the *dipole approximation*, by taking the field at the origin in the interaction hamiltonian,

$$H_I(t) = -ex(t) E(0, t) \qquad (2.2.9)$$

where $x(t)$ is the x-component of the position operator for the electron. This form of the interaction is equivalent to the usual one, obtained by the replacement $p \rightarrow p - eA$, in the dipole approximation[6].

We assume that the atom is exposed to the field (e.g., by removing a shutter) only at $t = 0$, the initial state of the system being

$$|g\psi_i\rangle = |g\rangle |\psi_i\rangle, \qquad (2.2.10)$$

where $|\psi_i\rangle$ is the initial state of the field. After $t = 0$, the system evolves according to the Schrödinger equation,

$$i\frac{\partial}{\partial t} |t\rangle = H_I |t\rangle, \qquad (2.2.11)$$

with

$$|t\rangle = U(t, 0) |0\rangle. \qquad (2.2.12)$$

The unitary evolution operator $U(t, 0)$ satisfies the integral equation

$$U(t, 0) = 1 - i \int_0^t H_I(t') U(t', 0) \, dt', \qquad (2.2.13)$$

which is equivalent to (2.2.11) plus the boundary condition $U(0, 0) = 1$.

The probability of finding the system in the state $|e\psi_f\rangle = |e\rangle\,|\psi_f\rangle$ at time t is given by

$$P_{g\psi_i \to e\psi_f} = |\langle e\psi_f|\,U(t, 0)\,|g\psi_i\rangle|^2. \tag{2.2.14}$$

In first-order perturbation theory, we have

$$U(t, 0) \approx 1 - i \int_0^t H_I(t')\,dt', \tag{2.2.15}$$

$$\langle e\psi_f|\,U(t, 0)\,|g\psi_i\rangle \approx -i \int_0^t dt'\,\langle e\psi_f|\,H_I(t')\,|g\psi_i\rangle$$

$$= ie \int_0^t dt'\,\langle e|\,x(t')\,|g\rangle\,\langle\psi_f|\,E(0, t')\,|\psi_i\rangle, \tag{2.2.16}$$

where we have used (2.2.9) and the relation $\langle e\,|\,g\rangle = 0$.

According to (2.2.8), we have

$$e\langle e|\,x(t')\,|g\rangle = e\langle e|\,e^{iH_0 t'}x(0)\,e^{-iH_0 t'}\,|g\rangle$$

$$= e\langle e|\,e^{iH_A t'}x(0)\,e^{-iH_A t'}\,|g\rangle = e^{i(\omega_e - \omega_g)t'}\,\langle e|\,ex\,(0)\,|g\rangle$$

$$= e^{i\omega_{eg}t'}d_{eg}, \tag{2.2.17}$$

where we have used the fact that H_A and $x(0)$ commute with the field hamiltonian H_F, and $d_{eg} = \langle e|\,ex(0)\,|g\rangle$ is the *transition dipole moment*, $\omega_{eg} = \omega_e - \omega_g$ the atomic transition frequency.

Substituting in (2.2.16), we get

$$\langle e\psi_f|\,U(t, 0)\,|g\psi_i\rangle = id_{eg} \int_0^t dt'\,e^{i\omega_{eg}t'}\,\langle\psi_f|\,E(0, t')\,|\psi_i\rangle. \tag{2.2.18}$$

We can now decompose $E(0, t')$ into positive and negative-frequency parts, according to (2.1.9). The privileged role played by $E^{(+)}$ in the absorption (photodetection) process then becomes apparent, for the contributions from $E^{(-)}$ lead to integrands of the form $\exp\,[i(\omega_{eg} + \omega_k)\,t']$, with $\omega_{eg} > 0$, $\omega_k > 0$, which are very small and rapidly oscillating, whereas those from $E^{(+)}$ contain factors of the form $\exp\,[i(\omega_{eg} - \omega_k)\,t']$, thus allowing energy to be conserved. For an observation made during a time interval t, energy is defined with an accuracy $\Delta\omega \sim t^{-1}$, but in practice this is always $\ll \omega_{eg}$ (mean optical periods are several orders of magnitude smaller than the resolving time of optical detectors), so that we can neglect the contribution from $E^{(-)}$.

The transition probability (2.2.14) becomes

$$P_{g\psi_i \to e\psi_f} = |d_{eg}|^2 \int_0^t dt' \int_0^t dt'' \, e^{i\omega_{eg}(t''-t')} \langle \psi_i | \, E^{(-)}(0, t') \, | \psi_f \rangle$$

$$\times \, \langle \psi_f | \, E^{(+)}(0, t'') \, | \psi_i \rangle. \tag{2.2.19}$$

We can now apply the same arguments of the previous section, summing over all possible final states of the field and taking a statistical average over initial states, to get [cf. (2.2.6)]

$$P_{g \to e}(t) = |d_{eg}|^2 \int_0^t dt' \int_0^t dt'' \, e^{i\omega_{eg}(t''-t')} G^{(1,1)}(0, t'; 0, t''). \tag{2.2.20}$$

So far we have assumed a discrete final state $|e\rangle$ for the electron; in the photoelectric effect, however, the electron makes a transition to a state in the continuous spectrum and (2.2.20) should be replaced by

$$p(t) = \int \varrho(\omega_{eg}) \, p_{g \to e}(t) \, d\omega_{eg}, \tag{2.2.21}$$

where $\varrho(\omega_{eg})$ is the density of final states per unit frequency interval. We should also take account of the fact that not all emitted photoelectrons are detected, by introducing an efficiency function of the detector.

Usually, the bandwidth $\delta\omega$ to which the detector is sensitive is much larger than the spectral width $\Delta\omega$ of the radiation field (broad band detector), so that the sensitivity of the detector may be taken as practically constant over the relevant spectral interval $\Delta\omega$; furthermore, $\delta\omega$ is also large as compared with t^{-1},

$$\delta\omega \gg \Delta\omega; \quad \delta\omega \gg t^{-1}. \tag{2.2.22}$$

Under these conditions we can effectively extend the integration in (2.2.21) over all frequencies, and use

$$\int_{-\infty}^{\infty} e^{i\omega_{eg}(t''-t')} \, d\omega_{eg} = 2\pi\delta(t'' - t'), \tag{2.2.23}$$

so that photon counts can be localized in time for a broad band detector. Taking into account (2.2.20) and (2.2.21), we get

$$p(t) = s \int_0^t G^{(1,1)}(0, t'; 0, t') \, dt', \tag{2.2.24}$$

where all factors contributing to the sensitivity of the detector have been lumped into the constant s.

The counting rate is obtained by differentiating with respect to t the total probability of photodetection in the interval $(0, t)$:

$$w(t) = \frac{dp(t)}{dt} = sG^{(1,1)}(0, t; 0, t).$$ (2.2.25)

Thus, an ideal photodetector allows us to measure $G^{(1,1)}(x, x)$ at a single spacetime point x, in agreement with the conclusion of the preceding section.

A real photodetector, of course, consists of a very large number of atoms. However, so long as the incident beam can be well approximated by a plane wave falling normally upon a thin photoelectric layer, a result of the above form remains true.

(c) Higher-order coherence functions

In order to motivate the definitions of higher-order coherence functions, let us consider[7] a "coincidence counting" experiment: n identical single-atom photon detectors are placed at the points $r_1, r_2, ..., r_n$ and exposed to the radiation field at $t = 0$. What is the probability $p^{(n)}(t)$ that *every* detector has absorbed just one photon from the field at time t?

In perturbation theory, the absorption of n photons corresponds to a term of order n in Dyson's well known[8] iterative solution of (2.2.13),

$$U(t, 0) = T\{\exp [-i \int_0^t H_I(t') \, dt']\} = \sum_{n=0}^{\infty} U^{(n)}(t, 0)$$

$$= \sum_{n=0}^{\infty} \frac{(-i)^n}{n!} \int_0^t dt_1 \, ... \int_0^t dt_n T[H_I(t_1) \, ... \, H_I(t_n)],$$ (2.2.26)

where T denotes the *time-ordered product*, defined by

$$T[\mathcal{O}(t_j) \mathcal{O}(t_k) \, ... \, \mathcal{O}(t_s)] = \mathcal{O}(t_1) \mathcal{O}(t_2) \, ... \, \mathcal{O}(t_n),$$

with

$$t_1 \geq t_2 \geq \cdots \geq t_n,$$ (2.2.27)

i.e., the operators are taken in the order of increasing time, read from right to left.

We assume that the atoms are far enough apart that their mutual interactions may be neglected (so that the atomic hamiltonians commute among themselves); in the dipole approximation, the interaction hamiltonian then becomes [cf. (2.2.9)]

$$H_I(t) = \sum_{j=1}^{n} H_{I,j}(t) = -e \sum_{j=1}^{n} x_j(t) E(r_j, t),$$ (2.2.28)

where the field is again assumed to be linearly polarized in the x-direction, and x_j is the x-component of the displacement of the jth electron from the fixed position $\mathbf{r}_j$.

Substituting (2.2.28) in (2.2.26) and expanding the products of inter-action hamiltonians, we obtain a large number of nth-order terms, many of which contain repetitions of the operator $H_{I,j}$ for the same atom j. Such terms are of no interest to us, since they correspond to the absorption of more than one photon by some atoms and none by others. Only terms in which each $H_{I,j}$ appears only once correspond to the process of interest. There are $n!$ such terms, all of them equal, because of the symmetry of (2.2.26) with respect to the atoms. Thus, we can specify one definite ordering of the $H_{I,j}$ and multiply the result by $n!$, so that the relevant part of $U^{(n)}(t, 0)$ is

$$(-i)^n \int_0^t dt_1 \ldots \int_0^t dt_n \, T[H_{I,1}(t_1) \ldots H_{I,n}(t_n)]$$

$$\rightarrow (ie)^n \int_0^t dt_1 \ldots \int_0^t dt_n \, T[x_1(t_1) \, E^{(+)}(\mathbf{r}_1, t_1) \ldots x_n(t_n) \, E^{(+)}(\mathbf{r}_n, t_n)], \quad (2.2.29)$$

where only the positive-frequency part gives an appreciable contribution, by the same argument already employed in the preceding section; each atom makes a transition from the ground state $|g\rangle$ to some excited state $|e_j\rangle$.

Since $E^{(+)}$ contains only annihilation operators [cf. (2.1.9)], it follows from (2.1.7) that the operators within square brackets in (2.2.29) commute among themselves, so that the time-ordered product may be replaced by an ordinary product.

It is clear, then, by analogy with (2.2.16), that if we take the squared modulus of the matrix element of (2.2.29) between initial and final states, summing over all possible final states (i.e., integrating over a band of final electronic states in the continuum), and taking the statistical average over initial states, the required probability $p^{(n)}(t)$ will be given by [cf. (2.2.24)]

$$p^{(n)}(t) = s^n \int_0^t dt_1' \ldots \int_0^t dt_n' \, G^{(n,n)}(\mathbf{r}_1, t_1', \ldots, \mathbf{r}_n, t_n'; \mathbf{r}_n, t_n', \ldots, \mathbf{r}_1, t_1'), \quad (2.2.30)$$

where we have assumed that all detectors are broad band with the same sensitivity s, and

$$G^{(n,n)}(x_1, \ldots, x_n; x_{n+1}, \ldots, x_{2n})$$

$$= \mathrm{Tr} \, [\varrho E^{(-)}(x_1) \ldots E^{(-)}(x_n) \, E^{(+)}(x_{n+1}) \ldots E^{(+)}(x_{2n})], \quad (x_j = \mathbf{r}_j, t_j). \quad (2.2.31)$$

If, instead of counting over the same time interval $(0, t)$ at each detector, we count up to time t_j at the jth detector (e.g., by closing a shutter at the appropriate times), we get, instead of (2.2.30),

$$p^{(n)}(t_1, ..., t_n) = s^n \int_0^{t_1} dt'_1 ... \int_0^{t_n} dt'_n G^{(n,n)}(\mathbf{r}_1, t'_1, ..., \mathbf{r}_n, t'_n;$$

$$\mathbf{r}_n, t'_n, ..., \mathbf{r}_1, t'_1) \tag{2.2.32}$$

as the probability of registering exactly one count at each detector during the corresponding time interval. The counting rate for n-fold (delayed) coincidences then is

$$w^{(n)}(t_1, ..., t_n) = \frac{\partial^n}{\partial t_1 ... \partial t_n} p^{(n)}(t_1, ..., t_n)$$

$$= s^n G^{(n,n)}(\mathbf{r}_1, t_1, ..., \mathbf{r}_n, t_n; \mathbf{r}_n, t_n, ..., \mathbf{r}_1, t_1). \tag{2.2.33}$$

This shows how one might measure, in principle, a special case of the *higher-order coherence function* (of order $2n$) defined by (2.2.31). We also see that, due to the special role played by absorption in the photodetection process, the coherence functions are associated with *normally-ordered products* of field operators (with all creation operators to the left and all annihilation operators to the right).

(d) Emission counters

In principle, it is also possible to have photon counters that operate by emission instead of absorption, although they may be difficult to realize in practice. Such a counter would operate by stimulated emission from excited atomic states due to the incident field. An example has been proposed by Mandel[9]: consider an atomic system having the level structure shown in Figure 2.1, with a long-lived metastable state m separated from the ground state g by a broad (short-lived) level e, the transition $m \to g$ being forbidden. The population of level m is initially raised by pumping (cf. § 5.1).

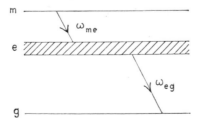

Figure 2.1 Level structure for emission counter. m: metastable state; e: broad excited level; g: ground state

Under the action of the incident field, assumed to have a suitable frequency spectrum, stimulated emission to level e takes place (corresponding to a photon of frequency ω_{me}), immediately followed by a transition to the ground state, with the emission of a photon of frequency ω_{eg}. We assume that ω_{eg} is sufficiently different from ω_{me} that we may discriminate between them and detect only the second photon; this allows us to detect the emission of the first photon indirectly, without having to absorb it.

Since only the emission part $E^{(-)}$ of the field plays a substantial role in stimulated emission, ideal counters of this type would allow us to measure the function

$$\mathrm{Tr}\,[\varrho E^{(+)}(x)\,E^{(-)}(x)], \tag{2.2.34}$$

and, more generally, functions of the type

$$\mathrm{Tr}\,[\varrho E^{(+)}(x_1)\,...\,E^{(+)}(x_n)\,E^{(-)}(x_{n+1})\,...\,E^{(-)}(x_{2n})], \tag{2.2.35}$$

corresponding to *anti-normally-ordered* products of field operators.

It is readily seen by applying the commutation relations (2.1.7), (2.1.8), that the counting rate (2.2.34) of an emission counter is higher than the corresponding rate (2.2.5) of a photodetector, the difference being independent of the incident field. This result has a very simple physical interpretation: in addition to stimulated emission from level m to e, there is also *spontaneous emission*, so that counts will be registered even in the absence of an external field. Thus, the difference between the counting rates of an emission counter and a photodetector is due to spontaneous emission.

2.3   GENERAL PROPERTIES OF COHERENCE FUNCTIONS

The discussion in § 2.2 suggests defining the general coherence function of order $n + m$ by[10]

$$G^{(n,m)}(x_1, ..., x_n; x_{n+1}, ..., x_{n+m})$$
$$= \mathrm{Tr}\,[\varrho E^{(-)}(x_1)\,...\,E^{(-)}(x_n)\,E^{(+)}(x_{n+1})\,...\,E^{(+)}(x_{n+m})], \tag{2.3.1}$$

where we have assumed, for simplicity, that the field is linearly polarized in a given direction (otherwise $G^{(n,m)}$ becomes a tensor of rank $n + m$ in the field components), and we use the notation (1.3.46).

In § 2.2 we considered only the special case $n = m$; in practice, this is the most important case, because one usually measures intensity correlations, in which the operators $E^{(-)}$, $E^{(+)}$ appear in pairs. Furthermore, for stationary fields, the frequency components of the field must satisfy some special relations[11] [generalizations of (1.3.32)] in order that $G^{(n,m)}$ be non-vanishing for $n \neq m$. Under these conditions, one can conceive of experiments that would allow one to measure such functions in some simple cases.

It is instructive to discuss the general properties of coherence functions by drawing a parallel with the well-known properties of Wightman functions[12]. Wightman functions are vacuum expectation values of products of field operators,

$$W(x_1, ..., x_n) = \langle 0| A(x_1) ... A(x_n) |0\rangle, \qquad (2.3.2)$$

where $|0\rangle$ denotes the vacuum state and we consider, for simplicity, a self-interacting neutral scalar field $A(x)$.

A vacuum expectation value is a special case of a statistical average, corresponding to the density operator

$$\varrho = |0\rangle \langle 0|. \qquad (2.3.3)$$

In this sense, Wightman functions are a special type of coherence functions. On the other hand, the physical motivation is quite different; in (2.3.2), it is the description of interacting fields that is of interest, whereas in (2.3.1) one usually (but not always!) deals with free radiation fields, and the interest is centered on the statistical properties of the field, embodied in its density operator ϱ. Because of the normal ordering, all the functions (2.3.1) would vanish identically in the vacuum state (2.3.3) [however, this would not be true for the functions (2.2.35)].

More generally, in an N-photon state, all functions $G^{(n,n)}$ vanish for $n > N$. This follows from the fact that, in such a state, $\varrho = |N\rangle \langle N|$, so that

$$E^{(+)}(x_1) ... E^{(+)}(x_n) \varrho = 0 \quad \text{for} \quad n > N. \qquad (2.3.4)$$

This remains true if we only know that the total number of photons does not exceed N, because, then, $\varrho = \sum_{n=1}^{N} p_n |n\rangle \langle n|$, where $|n\rangle$ denotes an n-photon state. Such states of the field, however, have no classical analogue, because their total energy goes to zero in the limit $\hbar \to 0$.

In spite of the differences between the physical problems of interest in connection with (2.3.1) and (2.3.2), the two sets of functions share a number of formal properties in common.

(I) HERMITICITY   For a neutral field $A(x)$, we have

$$W^*(x_1, ..., x_n) = W(x_n, ..., x_1). \qquad (2.3.5)$$

Similarly, since $E^{(-)}$ and $E^{(+)}$ are hermitian conjugates, (2.3.1) implies

$$G^{(n,m)*}(x_1, ..., x_n; x_{n+1}, ..., x_{n+m}) = G^{(m,n)}(x_{n+m}, ..., x_{n+1}; x_n, ..., x_1). \qquad (2.3.6)$$

(II) INVARIANCE PROPERTIES   By combining the transformation properties of a scalar field under an inhomogeneous Lorentz transformation $(a, \Lambda) x \to \Lambda x + a$ with the invariance of the vacuum state under such a transformation, one finds

$$W(x_1, ..., x_n) = W(\Lambda x_1 + a, ..., \Lambda x_n + a).$$

Similarly, if we assume that the density operator is invariant under some unitary transformation U,

$$U\varrho U^{-1} = \varrho, \tag{2.3.7}$$

so that it commutes with g, the infinitesimal generator of U,

$$[\varrho, g] = 0, \tag{2.3.8}$$

it follows from (2.3.1) that

$$G^{(n,m)}(x_1, ..., x_n; x_{n+1}, ..., x_{n+m}) = \text{Tr}\,[U\varrho U^{-1} E^{(-)}(x_1) ... E^{(+)}(x_{n+m})]$$

$$= \text{Tr}\,[\varrho U^{-1} E^{(-)}(x_1)\, UU^{-1} ... UU^{-1} E^{+}(x_{n+m})\, U],$$

$$= \text{Tr}\,[\varrho E_U^{(-)}(x_1) ... E_U^{(+)}(x_{n+m})], \tag{2.3.9}$$

where

$$E_U(x) = UE(x)\, U^{-1}. \tag{2.3.10}$$

Thus, for a *statistically homogeneous* and *time-stationary* field[13], we have invariance under spacetime translations $x \to x + a$, and we may take

$$U = e^{iP \cdot a}, \tag{2.3.11}$$

where P is the total four-momentum of the system; (2.3.9) becomes

$$G^{(n,m)}(x_1, ..., x_{n+m}) = G^{(n,m)}(x_1 + a, ..., x_{n+m} + a) \tag{2.3.12}$$

so that the coherence functions actually depend only on the $n + m - 1$ differences between their arguments, $\xi_i = x_{i+1} - x_i$.

If we have only *time-stationarity*, (2.3.12) is replaced by

$$G^{(n,m)}(\mathbf{r}_1, t_1, ..., \mathbf{r}_{n+m}, t_{n+m}) = G^{(n,m)}(\mathbf{r}_1, t_1 + \tau, ..., \mathbf{r}_{n+m}, t_{n+m} + \tau), \tag{2.3.13}$$

so that it depends only on the differences $\tau_i = t_{i+1} - t_i$ [cf. (1.3.14)].

As an example we may take *blackbody radiation*. The density operator for the radiation field in thermal equilibrium at temperature T is given by the expression[14]

$$\varrho = e^{-\beta H}/\text{Tr}\,(e^{-\beta H}), \quad \beta = 1/\varkappa T, \tag{2.3.14}$$

where $\varkappa$ is the Boltzmann constant and H is the Hamiltonian of the radiation field. Since H, and consequently, also, ϱ, commute with the total four-momentum, we conclude that blackbody radiation is time-stationary and statistically homogeneous; since H also commutes with the total angular momentum, it follows also that the radiation is statistically isotropic.

(III) LOCAL COMMUTATIVITY   For Wightman functions, the local commutation rules

$$[A(x), A(y)] = 0 \quad \text{for} \quad (x - y)^2 < 0, \tag{2.3.15}$$

imply

$$W(x_1, ..., x_n) = W(P(x_1, ..., x_n)), \tag{2.3.16}$$

where $P(x_1, ..., x_n)$ is any permutation such that $(x_i - x_j)^2 < 0$ for any (i, j) exchanged by P.

For coherence functions, it follows from (2.1.7) to (2.1.9) and (2.3.1) that

$$G^{(n,m)}(x_1, ..., x_n; x_{n+1}, ..., x_{n+m}) = G^{(n,m)}(P(x_1, ..., x_n); P'(x_{n+1}, ..., x_{n+m})), \tag{2.3.17}$$

where P and P' are any permutations. However, already for a free scalar field,

$$[A^{(+)}(x), A^{(-)}(y)] = i\Delta^{(+)}(x - y) \tag{2.3.18}$$

does *not* vanish for $(x - y)^2 < 0$, so that one may not permute any element in the first group of arguments in $G^{(n,m)}$ with any element in the second group without getting extra contributions, even if their separation is spacelike.

To see the origin of this difference and to show that no violation of causality is involved, let us consider the simple case of Young's interference experiment (Fig. 1.2), with pinholes at **x**, **y** and the detector placed symmetrically (equidistant from the pinholes). By analogy with (1.3.12), it is readily seen that a photodetector would allow us to measure the *spatial* coherence function Re $G^{(1,1)}(x; y) = $ Re Tr $[\varrho E^{(-)}(x) E^{(+)}(y)]$, with $x^0 = y^0$ (by symmetry), so that $(x - y)^2 < 0$. On the other hand, according to § 2.2(d), an emission counter placed at the same position would measure instead Re Tr $[\varrho E^{(+)}(y) E^{(-)}(x)]$. The difference between the two measurements, as we have seen in § 2.2(d), arises from the effect of *spontaneous emission*, which takes place at the location of the detector, so that no violation of causality is involved.

(IV) SPECTRAL CONDITION   For Wightman functions, if $\tilde{W}(p_1, ..., p_{n-1})$ is the Fourier transform of $W(\xi_1, ..., \xi_{n-1})$, where $\xi_i = x_{i+1} - x_i$ (using the invariance under spacetime translations, as in (2.3.12)), the spectral condition states that $\tilde{W}$ vanishes if any of the p_i does not lie in the energy-momentum spectrum for the states; thus, all p_i must lie on or within the forward light cone V_+. This implies that W is the boundary value of an analytic function $W(\xi_1 - i\eta_1, ..., \xi_{n-1} - i\eta_{n-1})$, holomorphic in the "future tube" $\eta_j \in V_+$.

For coherence functions, there is an analogous property for the Fourier decomposition in the *time domain*: by construction, $G^{(n,m)}(x_1, ..., x_n;$

$x_{n+1}, ..., x_{n+m})$ contains only positive frequencies in the m "destruction operator arguments", and only negative frequencies in the n "creation operator arguments". It follows that $G^{(n,m)}$ is the boundary value of an analytic function of the time variables, holomorphic in $\text{Im}\,(x_i^0) < 0$ for $n + 1 \leq i \leq n + m$, and in $\text{Im}\,(x_i^0) > 0$ for $1 \leq i \leq n$. In the particular case of $G^{(1,1)}$, this is the analogue of the property already discussed in § 1.3(c).

(V) POSITIVE-DEFINITENESS CONDITIONS   For Wightman functions, these conditions arise from the fact that an arbitrary state vector $|\psi\rangle$ has non-negative norm, $\|\,|\psi\rangle\|^2 \geq 0$. Taking

$$|\psi\rangle = f_0\,|0\rangle + \int d^4x\,f_1(x)\,A(x)\,|0\rangle$$

$$+ \int\int d^4x_1\,d^4x_2\,f_2(x_1, x_2)\,A(x_1)\,A(x_2)\,|0\rangle + \cdots, \qquad (2.3.19)$$

this leads to

$$\sum_{j,k} \int \cdots \int d^4x_1 \ldots d^4x_j f_j^*(x_1, ..., x_j)\,W(x_j, ..., x_1, y_1, ..., y_k)$$

$$f_k(y_1, ..., y_k)\,d^4y_1 \ldots d^4y_k \geq 0 \qquad (2.3.20)$$

for any choice of the test functions f_i; usually, one restricts the above series to a finite number of $f_i \neq 0$, to avoid convergence problems.

For coherence functions we start from the inequality

$$\text{Tr}\,(\varrho\mathcal{O}^+\mathcal{O}) \geq 0, \qquad (2.3.21)$$

valid for any operator $\mathcal{O}$, which is an immediate consequence of (2.1.15), (2.1.19). Taking

$$\mathcal{O} = \sum_j \int \cdots \int f_j(x_1, ..., x_j)\,E^{(+)}(x_1) \ldots E^{(+)}(x_j)\,d^4x_1 \ldots d^4x_j \qquad (2.3.22)$$

we find, by (2.3.1),

$$\sum_{j,k} \int \cdots \int d^4x_1 \ldots d^4x_j f_j^*(x_1, ..., x_j)\,G^{(j,k)}(x_j, ..., x_1; y_1, ..., y_k)$$

$$f_k(y_1, ..., y_k)\,d^4y_1 \ldots d^4y_k \geq 0 \qquad (2.3.23)$$

for any choice of the functions f_i.

By specializing the choice of the f_i, one immediately obtains some inequalities due to Glauber[15].

(a) $f_n(x_1, ..., x_n) = \delta(x_1 - \xi_1) \ldots \delta(x_n - \xi_n)$;   all other $f_i = 0$,

leads to

$$G^{(n,n)}(x_n, ..., x_1; x_1, ..., x_n) \geq 0, \qquad (2.3.24)$$

as was to be expected, in view of the physical interpretation of this quantity as an average intensity (for $n = 1$) or as an n-fold delayed coincidence counting rate (for $n > 1$).

(b) $f_m(x_1, \ldots, x_m) = \lambda_m \delta(x_1 - \xi_1) \ldots \delta(x_m - \xi_m);$

$f_n(x_1, \ldots, x_n) = \lambda_n \delta(x_1 - \eta_1) \ldots \delta(x_n - \eta_n);$ all other $f_i = 0$,

leads to

$|\lambda_m|^2 G^{(m,m)}(x_m, \ldots, x_1; x_1, \ldots, x_m) + \lambda_m^* \lambda_n G^{(m,n)}(x_m, \ldots, x_1; y_1, \ldots, y_n)$

$+ \lambda_m \lambda_n^* G^{(n,m)}(y_n, \ldots, y_1; x_1, \ldots, x_m) + |\lambda_n|^2 G^{(n,n)}(y_n, \ldots, y_1; y_1, \ldots, y_n) \geq 0,$

for any choice of the complex coefficients λ_m, λ_n. This requires

$$|G^{(m,n)}(x_m, \ldots, x_1; y_1, \ldots, y_n)|^2 \leq G^{(m,m)}(x_m, \ldots, x_1; x_1, \ldots, x_m)$$

$$\times G^{(n,n)}(y_n, \ldots, y_1; y_1, \ldots, y_n). \tag{2.3.25}$$

(c) $f_1(x) = \sum_{i=1}^{n} \lambda_i \delta(x - x_i),$ all other $f_i = 0,$

leads to

$$\sum_{i=1}^{n} \sum_{j=1}^{n} \lambda_i^* \lambda_j G^{(1,1)}(x_i; x_j) \geq 0. \tag{2.3.26}$$

In view of (2.3.6), this implies that $\|G^{(1,1)}(x_i; x_j)\|$ is a hermitian positive semi-definite matrix, so that[16]

$$\det \|G^{(1,1)}(x_i; x_j)\| \geq 0 \quad \text{for all} \quad i, j. \tag{2.3.27}$$

For $n = 1$ in (2.3.26), this corresponds to (2.3.24) for $n = 1$. For $n = 2$, it implies

$$|G^{(1,1)}(x_1, x_2)|^2 \leq G^{(1,1)}(x_1, x_1) G^{(1,1)}(x_2, x_2), \tag{2.3.28}$$

which may also be regarded as a special case of (2.3.25). This allows us to define, by analogy with (1.3.17), the normalized *complex degree of second-order coherence*

$$\gamma(x_1, x_2) = \frac{G^{(1,1)}(x_1, x_2)}{[G^{(1,1)}(x_1, x_1) G^{(1,1)}(x_2, x_2)]^{1/2}} ; \quad |\gamma| \leq 1. \tag{2.3.29}$$

If we take only $f_1 \neq 0$ in (2.3.23) and assume statistical homogeneity and time-stationarity, we get

$$\int \int d^4x \, f^*(x) \, G^{(1,1)}(x - y) f(y) \, d^4y \geq 0. \tag{2.3.30}$$

According to the *Bochner-Schwartz theorem*[17], (2.3.30) is a necessary and sufficient condition for $G^{(1,1)}(\xi)$ to be the Fourier transform of a positive tempered measure. This can readily be verified for the case (2.3.14) of blackbody radiation, in which[18]

$$\text{Tr}\,[\varrho E_i^{(-)}(x_1)\,E_i^{(+)}(x_2)] = G_{ii}^{(1,1)}(\xi = x_2 - x_1)$$

$$= \frac{1}{2\pi^2} \int d^4k\, e^{ik\cdot\xi} \theta(k_0)\, \delta(k^2)\, \frac{(|\mathbf{k}|^2 - k_i^2)}{(e^{\beta|\mathbf{k}|} - 1)}. \qquad (2.3.31)$$

If we have only time-stationarity, we may still apply this result in the time domain, by defining the *cross-spectral density*

$$W(\mathbf{x}_1, \mathbf{x}_2, \nu) = \int G^{(1,1)}(\mathbf{x}_1, \mathbf{x}_2, \tau = t_2 - t_1)\, e^{2\pi i\nu\tau}\, d\tau. \qquad (2.3.32)$$

Choosing $f(x) = \varphi(\mathbf{x})\,\psi(t)$ and applying the Bochner-Schwartz theorem to the analogue of (2.3.30), we find[19]

$$\int d^3x_1 \int d^3x_2\, \varphi^*(\mathbf{x}_2)\, W(\mathbf{x}_1, \mathbf{x}_2, \nu)\, \varphi(\mathbf{x}_1) \geqq 0. \qquad (2.3.33)$$

In particular, for $\varphi(\mathbf{x}) = \delta(\mathbf{x} - \xi)$, it follows that

$$W(\mathbf{x}, \mathbf{x}, \nu) \geqq 0, \qquad (2.3.34)$$

which, by (1.3.33), expresses the positive semi-definiteness of the spectral density.

A final point in this parallel discussion of the general properties of Wightman functions and coherence functions concerns the analogue of *Wightman's reconstruction theorem*, according to which a field theory is completely characterized by the set of all its Wightman functions.

The problem that might be regarded as analogous to this one in coherence theory would be: given the set of all coherence functions $G^{(n,m)}$ [verifying properties (I)–(V)], can one reconstruct the density operator of the system? We postpone to § 4.4 the discussion of this problem.

2.4   COHERENT FIELDS

(a) Second-order coherence

Similarly to (1.3.11), it follows from (2.2.5) that the counting rate of an ideal photodetector in Young's interference experiment would be proportional to (assuming a symmetrical arrangement, so that $K_1 = K_2$)

$$\text{Tr}\,\{\varrho[E^{(-)}(x_1) + E^{(-)}(x_2)]\,[E^{(+)}(x_1) + E^{(+)}(x_2)]\}$$

$$= G^{(1,1)}(x_1, x_1) + G^{(1,1)}(x_2, x_2) + 2\text{Re}\,G^{(1,1)}(x_1, x_2). \qquad (2.4.1)$$

Thus, with the degree of second-order coherence $\gamma(x_1, x_2)$ defined as in (2.3.29), the visibility of the interference fringes is still given by $|\gamma|$, as in

(1.3.27). The maximum visibility is attained when

$$|\gamma(x_1, x_2)| = 1. \tag{2.4.2}$$

We then have complete second-order coherence between the field at x_1 and that at x_2. It is useful to regard the field throughout space and time as a single dynamical system and to say that the *field* possesses second-order coherence when (2.4.2) is satisfied for *all* pairs of spacetime points (x_1, x_2). This is, of course, an idealization which can only be approximated in practice over a finite spacetime region, but it is nevertheless useful, like the concept of a plane wave.

According to (2.3.29), condition (2.4.2) is equivalent to

$$|G^{(1,1)}(x_1, x_2)|^2 = G^{(1,1)}(x_1, x_1)\, G^{(1,1)}(x_2, x_2). \tag{2.4.3}$$

This suggests that $G^{(1,1)}(x_1, x_2)$ may be *factorized* in the form

$$G^{(1,1)}(x_1, x_2) = A(x_1)\, B(x_2). \tag{2.4.4}$$

By (2.3.6), this would imply $A^*(x_1)\, B^*(x_2) = A(x_2)\, B(x_1)$, i.e., $A^*(x_1)/B(x_1) = A(x_2)/B^*(x_2) = \alpha$, where α is a real constant ≥ 0, as we see from (2.4.4) for $x_1 = x_2$ and (2.3.24) for $n = 1$. If we then introduce $V(x) = \sqrt{\alpha} B(x)$, (2.4.4) becomes

$$G^{(1,1)}(x_1, x_2) = V^*(x_1)\, V(x_2), \tag{2.4.5}$$

and $V(x)$ is determined up to a constant phase factor.

It will now be shown that *the factorization property (2.4.5) is not only sufficient, but also necessary for the field to possess second-order coherence*[20], i.e., for (2.4.2) to be satisfied.

In fact, let x_0 be an arbitrary spacetime point at which the intensity $G^{(1,1)}(x_0, x_0) \neq 0$, and consider the operator

$$\mathcal{O} = E^{(+)}(x) - \frac{G^{(1,1)}(x_0, x)}{G^{(1,1)}(x_0, x_0)}\, E^{(+)}(x_0). \tag{2.4.6}$$

Taking into account (2.3.1) and (2.3.6), we find

$$\mathrm{Tr}\,(\varrho \mathcal{O}^+ \mathcal{O}) = G^{(1,1)}(x, x) - \frac{|G^{(1,1)}(x_0, x)|^2}{G^{(1,1)}(x_0, x_0)} = 0, \tag{2.4.7}$$

as a consequence of (2.4.3), for any point x.

It then follows from (2.1.15) that

$$\sum_n p_n \langle \psi_n|\, \mathcal{O}^+ \mathcal{O}\, |\psi_n\rangle = \sum_n p_n \|\mathcal{O}|\psi_n\rangle\|^2 = 0,$$

i.e., $\mathcal{O}\,|\psi_n\rangle = 0$ for any n such that $p_n \neq 0$. Thus,

$$\mathcal{O}\varrho = \varrho \mathcal{O}^+ = 0, \tag{2.4.8}$$

3*

i.e., by (2.4.6),

$$E^{(+)}(x)\, \varrho = \frac{G^{(1,1)}(x_0, x)}{G^{(1,1)}(x_0, x_0)}\, E^{(+)}(x_0)\, \varrho, \tag{2.4.9}$$

and

$$\varrho E^{(-)}(x) = \varrho E^{(-)}(x_0) \frac{G^{(1,1)}(x, x_0)}{G^{(1,1)}(x_0, x_0)}. \tag{2.4.10}$$

This implies

$$G^{(1,1)}(x_1, x_2) = \mathrm{Tr}\,[\varrho E^{(-)}(x_1)\, E^{(+)}(x_2)] = \frac{G^{(1,1)}(x_1, x_0)\, G^{(1,1)}(x_0, x_2)}{G^{(1,1)}(x_0, x_0)}.$$

If we define

$$\tag{2.4.11}$$

$$V_{x_0}(x) = \frac{G^{(1,1)}(x_0, x)}{\sqrt{G^{(1,1)}(x_0, x_0)}}, \tag{2.4.12}$$

the result (2.4.11) becomes what we wanted to prove, i.e., that

$$G^{(1,1)}(x_1, x_2) = V_{x_0}^{*}(x_1)\, V_{x_0}(x_2). \tag{2.4.13}$$

Furthermore, the dependence of (2.4.12) on x_0 is only a trivial one; if we take any other reference point x_0' (with $G^{(1,1)}(x_0', x_0') \neq 0$), it follows from (2.4.11) with $x_1 = x_0'$, $x_2 = x$ that [cf. (2.3.29)]

$$V_{x'_0}(x) = \gamma(x_0', x_0)\, V_{x_0}(x), \tag{2.4.14}$$

so that, by (2.4.2), $V_{x'_0}(x)$ differs from $V_{x_0}(x)$ only by a phase factor; as we have seen in connection with (2.4.5), the choice of such a factor does not affect the coherence functions.

(b) Higher-order coherence

By analogy with (2.3.29), Glauber introduced a "degree of coherence" of order $2n$ by

$$g^{(n,n)}(x_1, \ldots, x_{2n}) = \frac{G^{(n,n)}(x_1, \ldots, x_{2n})}{\sqrt{G^{(1,1)}(x_1, x_1)} \cdots \sqrt{G^{(1,1)}(x_{2n}, x_{2n})}}, \tag{2.4.15}$$

and he defined as a *coherent field of order 2n* a field such that

$$|g^{(m,m)}(x_1, \ldots, x_{2m})| = 1, \quad 1 \leq m \leq n, \tag{2.4.16}$$

for any set of spacetime points $x_1, \ldots, x_{2n}$. This implies, in particular, that

$$G^{(m,m)}(x_1, \ldots, x_m; x_m, \ldots, x_1) = G^{(1,1)}(x_1, x_1) \ldots G^{(1,1)}(x_m, x_m)$$

$$(1 \leq m \leq n), \tag{2.4.17}$$

i.e., the m-fold coincidence-counting rate of m ideal photon detectors placed at $x_1, \ldots, x_m$ factors into the product of the counting rates that would be registered by each detector in the absence of the others: there are *no statistical correlations* between the counting rates of the detectors. It will be seen later [cf. § 4.3(b)] that the presence of such correlations is an indication of the existence of some random fluctuations ("noise") in the field; thus, coherence, in the above sense, corresponds to "noiselessness".

Since (2.4.16) must be valid, in particular, for $m = 1$, we may apply the results (2.4.9), (2.4.10) that follow from second-order coherence. They imply [with the help of (2.3.17)]

$$G^{(m,m)}(x_1, ..., x_{2m}) = \text{Tr}\,[\varrho E^{(-)}(x_1) \ldots E^{(-)}(x_m) E^{(+)}(x_{m+1}) \ldots E^{(+)}(x_{2m})]$$

$$= \frac{G^{(1,1)}(x_1, x_0)}{G^{(1,1)}(x_0, x_0)} \cdots \frac{G^{(1,1)}(x_m, x_0)}{G^{(1,1)}(x_0, x_0)} \text{Tr}\,[\varrho E^{(-)}(x_0) \ldots E^{(-)}(x_0) E^{(+)}(x_0) \ldots$$

$$\ldots E^{(+)}(x_0)] \times \frac{G^{(1,1)}(x_0, x_{m+1})}{G^{(1,1)}(x_0, x_0)} \cdots \frac{G^{(1,1)}(x_0, x_{2m})}{G^{(1,1)}(x_0, x_0)}, \qquad (2.4.18)$$

i.e., taking into account (2.4.12) and (2.4.15),

$$G^{(m,m)}(x_1, ..., x_{2m}) = g^{(m,m)}(x_0, ..., x_0)\, V^*(x_1) \ldots V^*(x_m)\, V(x_{m+1}) \ldots V(x_{2m}).$$
$$(2.4.19)$$

According to the discussion following (2.4.13), the dependence of $g^{(m,m)}$ on x_0 is only an apparent one; $g^{(m,m)}$ is a constant which, by (2.2.33), is proportional to the probability per unit (time)^m of detecting m photons at an arbitrary spacetime point with an ideal photodetector. Thus, merely from the fact that the field possesses second-order coherence, it already follows that *all higher-order coherence functions factorize* in the form (2.4.19); if, in addition, we know that the field possesses coherence of order $2n$, it follows from (2.4.16) and (2.3.24) that

$$g^{(m,m)} = 1, \qquad (2.4.20)$$

i.e.,

$$G^{(m,m)}(x_1, ..., x_{2m}) = V^*(x_1) \ldots V^*(x_m)\, V(x_{m+1}) \ldots V(x_{2m}), \quad 1 \leqq m \leqq n.$$
$$(2.4.21)$$

In this case, therefore, the factorization property (2.4.13) extends to coherence functions up to order $2n$, with the same function $V(x)$. Thus, if we have *full coherence* (to all orders), *all* coherence functions $G^{(m,m)}$ factorize in the form (2.4.21) in terms of a single function $V(x)$ (defined up to a constant phase factor).

The "degree of coherence" (2.4.15), unlike (2.3.29), does not satisfy the condition $0 \leqq |g^{(n,n)}| \leqq 1$ for $n > 1$. Alternative definitions of higher-order degrees of coherence have been proposed by Mehta[21] and Sudarshan[22]. However, in view of the scarcity of experimental data beyond the fourth order, a discussion of the relative merits of the various definitions seems largely academic at present.

(c) Density operators for coherent fields

A well-known example of a field possessing second-order coherence is a monochromatic field. In the plane-wave expansion (2.1.9), it would be associated with just a single term, but we can also consider a more general expansion in terms of any set of modes,

$$E^{(+)}(\mathbf{r}, t) = i \sum_k \sqrt{\frac{\omega_k}{2}} \, a_k u_k(\mathbf{r}) \exp\,(-i\omega_k t). \qquad (2.4.22)$$

If only one frequency ω_l is excited, the most general associated density operator in the Fock (occupation number) representation is of the form [cf. (2.1.25)]

$$\varrho = \sum_{i,j} \varrho_{ij} |i\rangle \langle j|, \qquad (2.4.23)$$

where $|i\rangle = (i!)^{-1/2}(a_l^+)^i |0\rangle$ denotes a state with i photons in the mode l. It follows that

$$G^{(1,1)}(\mathbf{r}_1, t_1; \mathbf{r}_2, t_2) = \frac{\omega_l}{2} u_l^*(\mathbf{r}_1)\, u_l(\mathbf{r}_2)\, \exp\,[-i\omega_l(t_2 - t_1)]$$

$$\times \sum_i \varrho_{ii} \langle i| a_l^+ a_l |i\rangle = V^*(x_1)\, V(x_2), \qquad (2.4.24)$$

where

$$V(x) = \left(\frac{\omega_l}{2} \sum_i i\varrho_{ii} \right)^{1/2} u_l(\mathbf{r})\, \exp\,(-i\omega_l t). \qquad (2.4.25)$$

Thus, when a single mode is excited, whether the field be in a pure state or in a mixture, the factorization condition for second-order coherence is satisfied.

A *stationary* field possessing second-order coherence is necessarily monochromatic. In fact, we must have in this case (omitting the spatial dependence)

$$G^{(1,1)}(t_2 - t_1) = V^*(t_1)\, V(t_2), \qquad (2.4.26)$$

and this functional equation, together with the fact that $G^{(1,1)}(\tau)$ must contain only positive frequencies, implies $V(t) \sim e^{-i\omega t}$, $\omega > 0$.

However, monochromaticity is by no means a necessary condition for second-order coherence. We may go over from the set of monochromatic modes employed in the expansion (2.4.22) to a different (nonmonochromatic) set by defining

$$v_m(\mathbf{r}, t) = i \sum_k c_{mk} \sqrt{\frac{\omega_k}{2}} \, u_k(\mathbf{r})\, \exp\,(-i\omega_k t), \qquad (2.4.27)$$

where the matrix $||c_{mk}||$ of the transformation coefficients is unitary,

$$\sum_m c^*_{mk'} c_{mk} = \delta_{kk'} = \sum_m c_{km} c^*_{k'm}. \tag{2.4.28}$$

In terms of the (generalized) modes v_m, (2.4.22) becomes

$$E^{(+)}(\mathbf{r}, t) = \sum_m b_m v_m(\mathbf{r}, t), \tag{2.4.29}$$

where

$$b_m = \sum_k c^*_{mk} a_k. \tag{2.4.30}$$

It follows from (2.4.28) that

$$[b_m, b_{m'}] = [b^+_m, b^+_{m'}] = 0; \quad [b_m, b^+_{m'}] = \delta_{m,m'}. \tag{2.4.31}$$

Thus [cf. (2.1.7)–(2.1.8)], we may interpret b^+_m and b_m as creation and annihilation operators for photons in the generalized mode v_m. If only this mode is excited, the density operator is again of the form (2.4.23), where $|i\rangle$ is a state with i photons in the mode v_m and (2.4.24) is replaced by

$$G^{(1,1)}(\mathbf{r}_1, t_1; \mathbf{r}_2, t_2) = v^*_m(\mathbf{r}_1, t_1) v_m(\mathbf{r}_2, t_2) \langle b^+_m b_m \rangle. \tag{2.4.32}$$

Thus, any (pure or mixed) state of a field in which only one mode (not necessarily monochromatic) is excited possesses second-order coherence. It was shown by Titulaer and Glauber[23] that the converse is also true, i.e., that *the most general type of field possessing second-order coherence may be regarded as one in which only a single mode (not necessarily monochromatic) is excited*; the corresponding density operator is of the form (2.4.23), where $|i\rangle$ is a state with i photons in the generalized mode in question. One may say that all photons in the field have identical wave packets [of the general form (2.4.27)]. This result is to be compared with the physical interpretation of the coherence volume given in § 1.2.

The concept of coherence, particularly in relation with the factorization of coherence functions, is closely related with the concept of "off-diagonal long-range order"[24], which plays an important role in the theories of superfluidity and superconductivity. The coherence functions introduced by Glauber represent an extension of the reduced density matrices employed in many-body theory. The relation between coherence and occupancy of a single mode discussed above may be compared with the notion of Bose-Einstein condensation into a single mode in superfluids. The Young interference patterns as a manifestation of coherence find their analogues for superconductors in the interference experiments performed with Josephson junctions[25].

The analogy between superfluids, superconductors and coherent modes of the electromagnetic field has been pointed out by several authors[26].

References

1. Cf., e.g., A. Messiah, *Quantum Mechanics*, vol. II, North-Holland Publishing Co., Amsterdam (1965) or J. D. Bjorken and S. D. Drell, *Relativistic Quantum Fields*, McGraw-Hill, New York (1965).
2. Cf. A. Messiah, loc. cit., p. 1032.
3. Cf. A. Messiah, *Quantum Mechanics*, vol. I, North-Holland Publishing Co., Amsterdam (1964), p. 331; K. Huang, *Statistical Mechanics*, John Wiley, New York (1963), chapter 9.
4. R. J. Glauber, *Phys. Rev.* **130**, 2529 (1963); *QOE* (see Ch. 1, Ref. 2), p. 65.
5. R. J. Glauber, *QOE*, p. 78.
6. Cf. R. G. Wolley, *Molec. Phys.* **22**, 1013 (1971); also, E. A. Power, *Introductory Quantum Electrodynamics*, Longmans, London (1964).
7. R. J. Glauber, *QOE*, p. 84.
8. Cf., e.g., P. Roman, *Advanced Quantum Theory*, Addison-Wesley, Reading (1965), p. 311.
9. L. Mandel, *Phys. Rev.* **152**, 438 (1966).
10. Note that this differs from Glauber's nomenclature; he considers mainly the case $m = n$, and he calls $G^{(n,n)}$ a coherence function of order n.
11. C. L. Mehta and L. Mandel, in *Electromagnetic Wave Theory*, edited by J. Brown, Pergamon Press, Oxford (1967), p. 1069; R. J. Glauber, *QO* (see Ch. 1, Ref. 3), p. 23.
12. A. S. Wightman, *Phys. Rev.* **101**, 860 (1956); R. F. Streater and A. S. Wightman, *PCT, Spin and Statistics, and All That*, W. Benjamin, New York (1964), p. 106.
13. Cf. J. H. Eberly and A. Kujawski, *Phys. Letters* **24**A, 426 (1967); D. Dialetis and C. L. Mehta, *Nuovo Cimento* **56**, 89 (1968).
14. Cf. A. Messiah, *Quantum Mechanics*, vol. I, North-Holland Publishing Co., Amsterdam (1964), pp. 337, 448.
15. R. J. Glauber, *QOE*, p. 90.
16. E. F. Beckenbach and R. Bellman, *Inequalities*, Springer-Verlag, Berlin (1965), p. 57.
17. I. M. Gel'fand and N. Ya. Vilenkin, *Generalized Functions*, Vol. 4, Academic Press, New York (1964), p. 157.
18. C. L. Mehta and E. Wolf, *Phys. Rev.* **134**, A 1149 (1964).
19. Cf. C. L. Mehta and E. Wolf, *Phys. Rev.* **157**, 1188 (1967).
20. U. M. Titulaer and R. J. Glauber, *Phys. Rev.* **140**, B 676 (1965).
21. C. L. Mehta, *J. Math. Phys.* **8**, 1798 (1967).
22. Cf. *KS* (see Ch. 1, Ref. 4), § 8.2.
23. U. M. Titulaer and R. J. Glauber, *Phys. Rev.* **145**, 1041 (1966).
24. O. Penrose and L. Onsager, *Phys. Rev.* **104**, 576 (1956); C. N. Yang, *Rev. Mod. Phys.* **34**, 694 (1962).
25. R. C. Jaklevic, J. Lambe, A. H. Silver, and J. E. Mercereau, *Phys. Rev. Letters* **12**, 159, 274 (1964).
26. P. C. Hohenberg and P. C. Martin, *Ann. Phys.* (*N. Y.*), **34**, 291 (1965); F. W. Cummings and J. R. Johnston, *Phys. Rev.* **151**, 105 (1966).

CHAPTER 3

Coherent States

3.1 DEFINITION OF COHERENT STATES

(a) Introduction

IN ORDER FOR a pure state $|V\rangle$ of the field to be fully coherent, we must have, according to (2.4.21),

$$\langle V|E^{(-)}(x_1) \ldots E^{(-)}(x_m) E^{(+)}(x_{m+1}) \ldots E^{(+)}(x_{2m})|V\rangle$$
$$= V^*(x_1) \ldots V^*(x_m) V(x_{m+1}) \ldots V(x_{2m}) \qquad (3.1.1)$$

for all values of m. The simplest way to fulfill this condition is for $|V\rangle$ to be an *eigenstate of the annihilation part of the field*,

$$E^{(+)}(x)|V\rangle = V(x)|V\rangle. \qquad (3.1.2)$$

In fact, this implies

$$\langle V|E^{(-)}(x) = V^*(x)\langle V|, \qquad (3.1.3)$$

and (3.1.1) follows from these relations. In fact they imply much more, namely, that

$$G^{(n,m)}(x_1, \ldots, x_n; x_{n+1}, \ldots, x_{n+m}) = V^*(x_1) \ldots V^*(x_n) V(x_{n+1}) \ldots V(x_{n+m})$$
$$(3.1.4)$$

also for $n \neq m$.

It may be shown[1] that, while (3.1.1) can be satisfied by more general states of the field, the states (3.1.2) are the only ones that satisfy the more restrictive condition (3.1.4), that has been proposed as an alternative definition of full coherence by some authors [in fact, the requirement that (3.1.4) be satisfied for (n, m) equal to $(0, 1)$ and $(1, 1)$ already leads uniquely to (3.1.2)]. The states (3.1.2) are called *coherent states*.

Let us expand $E^{(+)}(x)$, as in (2.4.22), in terms of an orthonormal set of modes $u_k(\mathbf{r})$,

$$\int u_k^*(\mathbf{r}) u_{k'}(\mathbf{r}) d^3r = \delta_{kk'}, \qquad (3.1.5)$$

and let the function $V(x)$ in (3.1.2) also be expanded in terms of this set,

$$V(\mathbf{r}, t) = i \sum_k \sqrt{\frac{\omega_k}{2}} v_k u_k(\mathbf{r}) \exp(-i\omega_k t). \qquad (3.1.6)$$

It then follows from (3.1.2), (2.4.22), (3.1.6) and the orthonormality of the modes that

$$a_k|V\rangle = v_k|V\rangle, \tag{3.1.7}$$

i.e., $|V\rangle$ must be an eigenvector of the annihilation operators a_k for all modes k [if the field is not linearly polarized, we must include polarization indices, as in (2.1.3)].

If we construct such eigenvectors for each mode,

$$a_k|v_k\rangle = v_k|v_k\rangle, \tag{3.1.8}$$

we can take $|V\rangle$ as a direct product

$$|V\rangle = |\{v_k\}\rangle = \prod_k |v_k\rangle, \tag{3.1.9}$$

where the label $\{v_k\}$ stands for the *sequence* of eigenvalues v_k for all modes.

The problem is thereby reduced to that of constructing coherent states for a single mode,

$$a|v\rangle = v|v\rangle, \tag{3.1.10}$$

where we have dropped the mode index k to simplify the notation.

(b) Coherent states for a single mode

It is well known[2] that a single mode of a free radiation field may be regarded as a dynamical system equivalent to a harmonic oscillator.

The vacuum state $|0\rangle$ (no photons in the mode) corresponds to the ground state of the oscillator,

$$a|0\rangle = 0, \quad \langle 0|0\rangle = 1, \tag{3.1.11}$$

and the normalized n-photon state $|n\rangle$ corresponds to the nth excited state of the oscillator, given by

$$|n\rangle = \frac{(a^+)^n}{\sqrt{n!}}|0\rangle, \quad \langle n|n\rangle = 1. \tag{3.1.12}$$

The commutation rules (2.1.7), (2.1.8) lead to

$$[a, (a^+)^n] = n(a^+)^{n-1}, \tag{3.1.13}$$

$$a|n\rangle = \sqrt{n}|n-1\rangle, \tag{3.1.14}$$

$$a^+|n\rangle = \sqrt{n+1}|n+1\rangle, \tag{3.1.15}$$

$$a^+a|n\rangle = n|n\rangle \quad \text{(number operator)}. \tag{3.1.16}$$

The canonical operators $Q(t)$, $P(t)$ of the associated harmonic oscillator, in the Heisenberg picture, are related with (a, a^+) by

$$Q(t) = q_0(ae^{-i\omega t} + a^+e^{i\omega t}), \tag{3.1.17}$$

$$P(t) = -i\omega q_0(ae^{-i\omega t} - a^+e^{i\omega t}), \tag{3.1.18}$$

$$a(t) = ae^{-i\omega t} = q_0[\omega Q(t) + iP(t)], \tag{3.1.19}$$

where

$$q_0 = \frac{1}{\sqrt{2\omega}} \tag{3.1.20}$$

is the square root of the "zero-point fluctuation" $\langle 0| Q^2 |0\rangle$ (in conventional units, $q_0 = (\hbar/2\omega)^{\frac{1}{2}}$).

Let us try to construct the coherent state $|v\rangle$ by means of an expansion in Fock (n-photon) states $|n\rangle$:

$$|v\rangle = \sum_{n=0}^{\infty} |n\rangle \langle n | v\rangle. \tag{3.1.21}$$

The expansion coefficients are given by [cf. (3.1.12), (3.1.10)]

$$\langle n | v\rangle = \langle 0| \frac{a^n}{\sqrt{n!}} |v\rangle = \frac{v^n}{\sqrt{n!}} \langle 0 | v\rangle, \tag{3.1.22}$$

so that

$$|v\rangle = \langle 0 | v\rangle \sum_{n=0}^{\infty} \frac{v^n}{\sqrt{n!}} |n\rangle.$$

The normalization factor $\langle 0 | v\rangle$ is determined by

$$1 = \langle v | v\rangle = |\langle 0 | v\rangle|^2 \sum_{n=0}^{\infty} \sum_{n'=0}^{\infty} \frac{v^{*n'}v^n}{\sqrt{n!n'!}} \langle n' | n\rangle$$

$$= |\langle 0 | v\rangle|^2 \sum_{n=0}^{\infty} \frac{|v|^{2n}}{n!} = \exp(|v|^2) |\langle 0 | v\rangle|^2$$

so that, by suitably choosing the phase,

$$\langle 0 | v\rangle = \exp(-\tfrac{1}{2} |v|^2), \tag{3.1.23}$$

and

$$|v\rangle = \exp(-\tfrac{1}{2} |v|^2) \sum_{n=0}^{\infty} \frac{v^n}{\sqrt{n!}} |n\rangle, \qquad \langle v | v\rangle = 1. \tag{3.1.24}$$

The eigenvalues v are all complex numbers. Since a is not hermitian, there is nothing surprising in its having complex eigenvalues. In fact, it might very well have turned out not to have any normalizable right eigenvectors at all; this happens with a^+, as may easily be verified; the existence of eigenvectors for a has been explicitly verified by constructing them. Note that the vacuum state (3.1.11) is also a coherent state.

According to (3.1.24), the number of photons that may be found in a coherent state is unbounded: it may range from 0 to ∞. The *average* number of photons in the state $|v\rangle$ is given by

$$\langle n \rangle = \langle v| \, a^+ a \, |v\rangle = |v|^2. \tag{3.1.25}$$

The probability of finding n photons in $|v\rangle$ is [cf. (3.1.22), (3.1.23)]

$$|\langle n \mid v\rangle|^2 = \frac{|v|^{2n}}{n!} \exp\left(-|v|^2\right) = \frac{\langle n\rangle^n}{n!} e^{-\langle n\rangle}, \tag{3.1.26}$$

which corresponds to a *Poisson distribution*.

According to (3.1.25), the modulus of the eigenvalue v has a simple physical interpretation: its square gives the average number of photons in $|v\rangle$. What about the phase of v? Let

$$v = |v| \, e^{i\varphi}. \tag{3.1.27}$$

Then, according to (3.1.17),

$$\langle v| \, Q(t) \, |v\rangle = 2q_0 \, |v| \cos{(\omega t - \varphi)}, \tag{3.1.28}$$

i.e., the expectation value of $Q(t)$ in a coherent state behaves precisely like the coordinate of a classical harmonic oscillator with amplitude proportional to $|v|$, and φ is the analogue of the oscillator phase.

The result (3.1.28) already indicates that coherent states should be suitable for discussing the approach to the classical limit. The correspondence principle may be expected to apply to states with large average number of photons, $\bar{n} \gg 1$, so that, by (3.1.25), $|v| \gg 1$. However, (3.1.28) remains valid even for small $|v|$, so that a formal analogy with classical results may sometimes be extended beyond the correspondence limit, as will be seen later.

Note that Fock states are not suitable for discussing the classical limit, because, by (3.1.14), (3.1.15), (3.1.17), (3.1.18),

$$\langle n| \, Q(t) \, |n\rangle = \langle n| \, P(t) \, |n\rangle = 0. \tag{3.1.29}$$

This is related with the fact that photon number and phase are to some extent complementary variables[3], so that the phase is undetermined when the photon number is fixed. As has already been mentioned in § 2.3, the energy of an n-photon state goes to zero in the classical limit ($\hbar \to 0$).

3.2 PROPERTIES OF COHERENT STATES

The coherent states have several remarkable properties that will now be discussed.

(a) Coherent states as displaced harmonic oscillator states

Substituting (3.1.12) in (3.1.24), we find

$$|v\rangle = \exp\left(-\tfrac{1}{2}|v|^2 + va^+\right)|0\rangle. \tag{3.2.1}$$

Since $\langle v\,|\,v\rangle = \langle 0\,|\,0\rangle = 1$, this suggests that the exponential operator in (3.2.1) is equivalent to a unitary operator. Consider the operator

$$U(v) = \exp\left(va^+ - v^*a\right).$$

It is unitary,

$$U^+(v) = U(-v) = [U(v)]^{-1}. \tag{3.2.2}$$

To show that it is equivalent to (3.2.1), we have to apply the following lemma[4], which is a special case of the Baker-Hausdorff theorem:

Lemma: If the commutator C of two operators A and B commutes with them,

$$C = [A, B], \quad [A, C] = [B, C] = 0, \tag{3.2.3}$$

we have

$$e^{A+B} = e^A e^B e^{-C/2} \tag{3.2.4}$$

Proof: Let

$$f(x) = e^{xA} e^{xB}.$$

Then,

$$\frac{df}{dx} = (A + e^{xA}Be^{-xA})f(x).$$

But

$$e^{xA}Be^{-xA} = B + x[A, B] + \frac{x^2}{2!}[A, [A,B]] + \frac{x^3}{3!}[A, [A, [A,B]]] + \cdots, \tag{3.2.5}$$

as may readily be seen by evaluating the coefficients in this Taylor series by the standard formulas. In view of (3.2.3), this becomes

$$e^{xA}Be^{-xA} = B + xC,$$

and the above result becomes

$$\frac{df}{dx} = (A + B + xC)f(x),$$

or, since $f(0) = 1$ and C commutes with $A + B$,

$$f(x) = e^{(A+B)x}e^{Cx^2/2}.$$

Taking $x = 1$, the result is equivalent to (3.2.4).

Now take $A = va^+$, $B = -v^*a$, so that $C = |v|^2$, and apply (3.2.4). The result is

$$U(v) = \exp(va^+ - v^*a) = \exp(va^+)\exp(-v^*a)\exp(-|v|^2/2), \quad (3.2.6)$$

so that

$$U(v)|0\rangle = \exp(-|v|^2/2)\exp(va^+)|0\rangle.$$

Comparing this with (3.2.1), we get

$$|v\rangle = U(v)|0\rangle. \quad (3.2.7)$$

By (3.2.2), it follows that

$$U(-v)|v\rangle = |0\rangle, \quad (3.2.8)$$

so that $U(v)$ and $U(-v)$ behave like creation and annihilation operators of coherent states.

We can also see this directly, by noting that [cf. (3.1.13)]

$$\exp(|v|^2/2)[a, U(v)] = [a, \exp(va^+)]\exp(-v^*a)$$

$$= \sum_{n=0}^{\infty} \frac{v^n}{n!}[a, (a^+)^n]\exp(-v^*a)$$

$$= v\sum_{n=1}^{\infty} \frac{(va^+)^{n-1}}{(n-1)!}\exp(-v^*a) = v\exp(|v|^2/2)\,U(v),$$

$$(3.2.9)$$

from which (3.2.7) follows by applying both sides to the vacuum state. This result may be rewritten as

$$U^{-1}(v)\,aU(v) = a + v, \quad (3.2.10)$$

and it follows that

$$U^{-1}(v)\,a^+U(v) = a^+ + v^*. \quad (3.2.11)$$

Thus, $U(v)$ acts on a as a *displacement operator* by the complex quantity v. By (3.1.17) and (3.1.27), this implies

$$U^{-1}(v)\,Q(t)\,U(v) = Q(t) + 2q_0|v|\cos(\omega t - \varphi). \quad (3.2.12)$$

The vacuum expectation value of this relation reproduces (3.1.28).

Let

$$Q(0) = Q, \quad P(0) = P, \quad ([Q, P] = i), \quad (3.2.13)$$

which may be regarded as the position and momentum operators in the Schrödinger picture. By analogy with (3.1.19), let us write

$$v = q_0(\omega q + ip), \quad (3.2.14)$$

so that the coherent state $|v\rangle$ may also be labeled by the real and imaginary parts of v as

$$|v\rangle = |q, p\rangle. \quad (3.2.15)$$

Substituting (3.1.19) for $t = 0$ into (3.2.6), and taking into account (3.2.14) and (3.1.20), we get

$$U(v) = U(q, p) = e^{i(pQ - qP)}. \qquad (3.2.16)$$

By (3.2.3), (3.2.4) and (3.2.13), this may be rewritten as

$$U(v) = e^{ipQ}e^{-iqP}e^{-ipq/2}. \qquad (3.2.17)$$

We can now find the wave function of a coherent state in the x-representation,

$$Q|x\rangle = x|x\rangle. \qquad (3.2.18)$$

It is given by

$$\langle x \mid v \rangle = \langle x| U(v) |0\rangle = e^{-ipq/2} \langle x| e^{ipx}e^{-iqP}|0\rangle$$

$$= e^{ipx - ipq/2} \langle x - q \mid 0\rangle = Ne^{ipx - ipq/2} \exp\left[-\left(\frac{x - q}{2q_0}\right)^2\right], \qquad (3.2.19)$$

where N is a normalization factor for the well-known Gaussian ground-state wave function $\langle x \mid 0 \rangle$ of the harmonic oscillator, and we have employed the fact that e^{-iqP} is a displacement operator. This shows explicitly that the coherent state wave function is just the Gaussian wave packet for the ground state of the harmonic oscillator, with its center displaced by q (and a phase factor corresponding to momentum p, so that the ground state momentum is displaced by p). The wave function in momentum space (Fourier transform) is also a Gaussian, centered at p.

According to (3.1.19) and (3.1.28), the wave packet at time t in the Schrödinger picture differs from that at time zero only by the fact that its center is displaced to a position given by (3.1.28), i.e., it just oscillates following the classical trajectory, without changing its shape in the course of time, as other Schrödinger wave packets would normally do. Wave packets of this kind were first considered by Schrödinger[5]. Their application to the quantum theory of coherence was advocated by Glauber[6].

(b) Coherent states as minimum-uncertainty states

According to the above discussion, the uncertainties ΔQ and ΔP for a coherent state are time-independent, so that we may compute them at $t = 0$. By (3.1.17),

$$\langle v| Q |v\rangle = q_0(v + v^*),$$

$$\langle v| Q^2 |v\rangle = q_0^2 \langle v| (a^2 + aa^+ + a^+a + a^{+2}) |v\rangle = q_0^2[(v + v^*)^2 + 1],$$

so that

$$(\Delta Q)^2 = \langle Q^2\rangle - \langle Q\rangle^2 = q_0^2. \qquad (3.2.20)$$

In agreement with the above discussion, this is independent of v and coincides with the zero-point fluctuation (3.1.20).

Similarly, by (3.1.18),

$$(\Delta P)^2 = \langle v| \, P^2 \, |v\rangle - \langle v| \, P \, |v\rangle^2 = -\omega^2 q_0^2 \, \langle v| \, (a^2 - aa^+ - a^+a + a^{+2}) \, |v\rangle$$
$$+ \omega_0^2 q_0^2 \, \langle v| \, (a - a^+) \, |v\rangle^2 = -\omega^2 q_0^2 [(v - v^*)^2 - 1] + \omega_0^2 q_0^2 (v - v^*)^2,$$

so that

$$(\Delta P)^2 = \omega_0^2 q_0^2. \tag{3.2.21}$$

Combining (3.2.20) and (3.2.21), we get

$$\Delta Q \Delta P = \omega_0 q_0^2 = \tfrac{1}{2}. \tag{3.2.22}$$

According to Heisenberg's uncertainty relation, we must have

$$\Delta Q \Delta P \geqq \tfrac{1}{2}. \tag{3.2.23}$$

Thus, according to (3.2.22), coherent states are *minimum-uncertainty states*. This agrees with the previous observation that they are as close as possible to classical states, and that coherence may be interpreted as "minimum noise".

The most general minimum-uncertainty wave packet[7] is given by (3.2.19), where, however, q_0 is an arbitrary real constant, not necessarily related to the oscillator frequency by (3.1.20). However, the coherent states are the only minimum-uncertainty wave packets which remain so in the course of time, while $\langle Q(t) \rangle$ and $\langle P(t) \rangle$ follow the classical harmonic oscillator trajectory.

(c) Coherent states and the forced harmonic oscillator

It follows from (3.2.19) that the coherent state $|p, q\rangle$ is an eigenstate of the hamiltonian

$$H = \tfrac{1}{2}[(P - p)^2 + \omega^2(Q - q)^2]$$
$$= \tfrac{1}{2}(P^2 + \omega^2 Q^2) - pP - \omega^2 qQ + \tfrac{1}{2}(p^2 + \omega^2 q^2). \tag{3.2.24}$$

More generally, we can let p and q be time dependent; the coherent state $|p(t), q(t)\rangle$ will still be an eigenstate, for each t, of the (non-conservative) hamiltonian

$$H(t) = \tfrac{1}{2}(P^2 + \omega^2 Q^2) - p(t)\, P - \omega^2 q(t)\, Q. \tag{3.2.25}$$

In particular, for $p(t) = 0$, this hamiltonian describes an oscillator subject to an external (c-number) driving force[8] $\omega^2 q(t)$. Such a force therefore preserves the coherence of an initially coherent state; in particular, if we start from the ground (vacuum) state, which is a special case of a coherent state, the final state after the application of the force is also a coherent state.

More generally, it can be shown[9] that, for a system of harmonic oscillators described by the operators a_k, a_k^+, all hamiltonians of the form (of which (3.2.25) is a particular case)

$$H(t) = \sum_{k,k'} \omega_{kk'}(t)\, a_k^+ a_{k'} + \sum_k [F_k(t)\, a_k^+ + F_k^*(t)\, a_k] + h(t), \qquad (3.2.26)$$

where $\omega_{kk'} = \omega_{k'k}^*$, F_k and $h = h^*$ are arbitrary functions of time, preserve the coherent character of an initially coherent state. This is the most general class of hamiltonians possessing this property.

(d) Non-orthogonality and overcompleteness

The scalar product of two coherent state vectors may readily be computed from (3.1.24):

$$\langle v' \mid v \rangle = \exp\left[-\tfrac{1}{2}(|v|^2 + |v'|^2)\right] \sum_{n,n'} \frac{(v'^*)^{n'} v^n}{\sqrt{n'!\,n!}}\, \delta_{n,n'}$$

$$= \exp\left(-\tfrac{1}{2}|v'|^2 + v'^*v - \tfrac{1}{2}|v|^2\right). \qquad (3.2.27)$$

It follows that
$$|\langle v' \mid v \rangle|^2 = \exp\left(-|v' - v|^2\right). \qquad (3.2.28)$$

Thus, two coherent states corresponding to different eigenvalues v and v' are *not* orthogonal. However, their overlap, according to (3.2.28), becomes very small when $|v - v'| \gg 1$, i.e., when the distance between the representative points in the complex v-plane is large compared with unity. If we think of the v-plane as a kind of "phase space" for the oscillator, each point v, according to (3.2.23), actually corresponds to an area of order unity (or $\hbar$, in conventional units) in phase space, and (3.2.28) becomes appreciable when the representative areas overlap significantly.

In spite of their non-orthogonality, one can expand an arbitrary Hilbert space vector in terms of the coherent states: actually, the set of all coherent states $|v\rangle$ (where v ranges over the whole complex plane) is an *overcomplete set*, i.e., subsets of this set are already complete (if we take the whole set of coherent states, they are not linearly independent, as will be seen below).

The analogue of the "completeness sum" for coherent states is the integral

$$J = \int |v\rangle \langle v|\, d^2v \qquad (3.2.29)$$

extended over the whole complex v-plane [cf. (1.3.8)]. By (3.1.24),

$$J = \sum_{n,n'} \frac{|n'\rangle \langle n|}{\sqrt{n'!\,n!}} \int (v^*)^{n'} v^n \exp\left(-|v|^2\right) d^2v.$$

Taking polar coordinates,

$$v = re^{i\theta}, \quad d^2v = r\,dr\,d\theta,$$ (3.2.30)

we find

$$J = \sum_{n,n'} \frac{|n'\rangle\langle n|}{\sqrt{n'!n!}} \int_0^\infty r^{n'+n+1} \exp(-r^2)\,dr \int_0^{2\pi} e^{i(n-n')\theta}\,d\theta$$

$$= 2\pi \sum_n \frac{|n\rangle\langle n|}{n!} \int_0^\infty r^{2n+1}e^{-r^2}\,dr = \pi \sum_n |n\rangle\langle n| = \pi.$$

Thus, finally,

$$\frac{1}{\pi} \int |v\rangle\langle v|\,d^2v = 1.$$ (3.2.31)

This is the resolution of unity ("completeness relation") for coherent states. It allows one to expand state vectors and operators in terms of co-herent states, as will now be seen.

3.3   REPRESENTATIONS IN TERMS OF COHERENT STATES

(a) Representation of state vectors

Let us consider an arbitrary Hilbert space vector $|f\rangle$, $\||f\rangle\| < \infty$. According to (3.2.31), it may be expanded in terms of coherent states by

$$|f\rangle = \frac{1}{\pi} \int d^2v \,\langle v|f\rangle\,|v\rangle$$ (3.3.1)

Let us also consider the expansion of $|f\rangle$ in Fock states,

$$|f\rangle = \sum_{n=0}^\infty f_n|n\rangle, \quad f_n = \langle n|f\rangle.$$ (3.3.2)

According to (3.1.22), (3.1.23), we have

$$\langle v|f\rangle = \exp(-\tfrac{1}{2}|v|^2) \sum_{n=0}^\infty f_n \frac{(v^*)^n}{\sqrt{n!}} = \exp(-\tfrac{1}{2}|v|^2)f(v^*),$$ (3.3.3)

where

$$f(z) = \sum_{n=0}^\infty f_n \frac{z^n}{\sqrt{n!}}.$$ (3.3.4)

We have

$$|f(z)| \leq \sum_{n=0}^\infty |f_n| \frac{|z|^n}{\sqrt{n!}} \leq \||f\rangle\| \sum_{n=0}^\infty \frac{|z|^n}{\sqrt{n!}},$$ (3.3.5)

where the last series converges for all values of $|z|$. In fact, the ratio of the $(n + 1)^{th}$ term to the nth is $|z|/\sqrt{n + 1}$, which $\to 0$ as $n \to \infty$. Thus, $f(z)$ is an *entire analytic function* of the complex variable z. It is not an arbitrary entire function, because its growth is restricted by (3.3.5),

$$|f(z)| \leq \| |f\rangle \| \exp\left(\tfrac{1}{2} |z|^2\right), \tag{3.3.6}$$

so that it is an entire function of *order*[10] ≤ 2.

Substituting (3.3.3) in (3.3.1), we find

$$|f\rangle = \frac{1}{\pi} \int d^2v \exp\left(-\tfrac{1}{2} |v|^2\right) f(v^*) |v\rangle, \tag{3.3.7}$$

which is the coherent-state expansion of $|f\rangle$. Conversely, if (3.3.7) is given, we may evaluate the expansion coefficients by taking the scalar product with $\langle v'|$ and using (3.2.27):

$$\langle v' | f\rangle = \frac{1}{\pi} \exp\left(-\tfrac{1}{2} |v'|^2\right) \int \exp\left(v'^*v - |v|^2\right) f(v^*) \, d^2v. \tag{3.3.8}$$

By means of the change of variable (3.2.30), we find

$$\frac{1}{\pi} \int \exp\left(v'^*v - |v|^2\right)(v^*)^n \, d^2v = \frac{1}{\pi} \int_0^\infty r^{n+1} \exp\left(-r^2\right) dr$$

$$\times \int_0^{2\pi} \exp\left(v'^*re^{i\theta}\right) e^{-in\theta} \, d\theta = \frac{1}{\pi i} \int_0^\infty r^{n+1} \exp\left(-r^2\right) dr \oint_{|w|=1} \exp\left(v'^*rw\right)$$

$$\times \frac{dw}{w^{n+1}} = 2 \frac{(v'^*)^n}{n!} \int_0^\infty \exp\left(-r^2\right) r^{2n+1} \, dr = (v'^*)^n. \tag{3.3.9}$$

If we assume that $f(v^*)$ is an entire function of v^* (and therefore can be expanded into an absolutely convergent power series), it follows from (3.3.9) that

$$\frac{1}{\pi} \int \exp\left(v'^*v - |v|^2\right) f(v^*) \, d^2v = f(v'^*), \tag{3.3.10}$$

and, therefore, by (3.3.8),

$$f(v^*) = \exp\left(\tfrac{1}{2} |v|^2\right) \langle v | f\rangle, \tag{3.3.11}$$

in agreement with (3.3.3). Thus, *if we make the above assumption*, the expansion coefficients in (3.3.7) are *uniquely determined*.

4*

On the other hand, if we allow the expansion coefficients in (3.3.7) to be arbitrary functions of v^* *and* v, this is by no means true. To see this, it suffices to note that

$$\int |v\rangle \exp\left(-\tfrac{1}{2}|v|^2\right) v^m \, d^2v = \sum_{n=0}^{\infty} \frac{|n\rangle}{\sqrt{n!}} \int \exp\left(-|v|^2\right) v^{m+n} \, d^2v$$

$$= \sum_{n=0}^{\infty} \frac{|n\rangle}{\sqrt{n!}} \int_0^{\infty} \exp\left(-r^2\right) r^{m+n+1} \, dr \int_0^{2\pi} e^{i(m+n)\theta} \, d\theta = 0 \quad (m \neq 0), \quad (3.3.12)$$

which is an example of a linear dependence relation among the states $|v\rangle$, illustrating their overcompleteness. At the same time, this implies that the right-hand side of (3.3.7) is unchanged if we substitute

$$f(v^*) \to g(v, v^*) = f(v^*) + \sum_{m \geq 1} c_m v^m. \quad (3.3.13)$$

Thus, overcompleteness leads to infinitely many equivalent expansions for the same state vector. Another example of a linear dependence relation is obtained by expanding a coherent state itself in terms of the others,

$$|v\rangle = \frac{1}{\pi} \int |v'\rangle \langle v'|v\rangle \, d^2v' = \frac{1}{\pi} \exp\left(-\tfrac{1}{2}|v|^2\right) \int \exp\left(-\tfrac{1}{2}|v'|^2 + v'^*v\right)|v'\rangle \, d^2v'.$$

$$(3.3.14)$$

However, with the additional requirement that $f(v^*)$ in (3.3.7) be an entire function of v^*, the expansion becomes unique. For a bra $\langle g|$, we have, similarly,

$$\langle g| = \frac{1}{\pi} \int d^2v \exp\left(-\tfrac{1}{2}|v|^2\right) [g(v^*)]^* \langle v|, \quad (3.3.15)$$

where $g(z) = \sum_{n=0}^{\infty} \langle n|g\rangle z^n/\sqrt{n!}$, as in (3.3.4).

The scalar product of two vectors is represented by

$$\langle g|f\rangle = \frac{1}{\pi^2} \int\int d^2v \, d^2v' \exp\left[-\tfrac{1}{2}(|v|^2 + |v'|^2)\right] [g(v^*)]^* f(v'^*) \langle v|v'\rangle$$

$$= \frac{1}{\pi^2} \int d^2v \exp\left(-|v|^2\right) [g(v^*)]^* \int \exp\left(v^*v' - |v'|^2\right) f(v'^*) \, d^2v',$$

and, with the help of (3.3.10), this becomes

$$\langle g|f\rangle = \frac{1}{\pi} \int [g(v^*)]^* f(v^*) \exp\left(-|v|^2\right) d^2v. \quad (3.3.16)$$

Through (3.3.7), (3.3.11) and (3.3.15) is defined a one-to-one correspondence between Hilbert space vectors and entire analytic functions, in which the scalar product is represented through (3.3.16). The *Hilbert space of analytic functions* thus defined was introduced and has been thoroughly investigated by Bargmann[11].

(b) Representation of operators

Let F be an arbitrary operator with the Fock representation

$$F = \sum_{m,n} |m\rangle F_{mn} \langle n|, \qquad F_{mn} = \langle m| F |n\rangle. \tag{3.3.17}$$

The corresponding representation in terms of coherent states is obtained with the help of (3.2.31):

$$F = \frac{1}{\pi^2} \int d^2v' \int d^2v\, |v'\rangle \langle v'| F |v\rangle \langle v|. \tag{3.3.18}$$

Substituting (3.3.17) and taking into account (3.1.22), (3.1.23), we get

$$\langle v'| F |v\rangle = \sum_{m,n} F_{mn} \langle v' | m\rangle \langle n | v\rangle$$

$$= \exp\left[-\tfrac{1}{2}(|v'|^2 + |v|^2)\right] F(v'^*, v), \tag{3.3.19}$$

where

$$F(v'^*, v) = \sum_{m=0}^{\infty} \sum_{n=0}^{\infty} F_{mn} \frac{(v'^*)^m v^n}{\sqrt{m!n!}}. \tag{3.3.20}$$

Substituting (3.3.19) in (3.3.18), we finally get the coherent-state expansion

$$F = \frac{1}{\pi^2} \int d^2v' \int d^2v \exp\left[-\tfrac{1}{2}(|v'|^2 + |v|^2)\right] F(v'^*, v) |v'\rangle \langle v|, \tag{3.3.21}$$

which may be regarded as the analogue of (3.3.7) for an operator.

If F is a *bounded operator* [cf. § 3.4(b)], so that F_{mn} is bounded, it follows, as for (3.3.5), that $F(v'^*, v)$, as defined by (3.3.20), is an entire analytic function of both complex variables. This is true even for a broad class of unbounded operators, e.g., those for which $F_{mn} = O(m^\alpha n^\beta)$, where α and β are fixed positive constants; they include, for example, the operators in (3.1.14)–(3.1.16). The requirement of analyticity in v'^* and v leads uniquely to the expansion coefficients (3.3.19).

A particularly important bounded operator to which (3.3.21) may be applied is the density operator, for which

$$\text{Tr}\,(\varrho^2) = \sum_{m,n} |\varrho_{m,n}|^2 \leq 1. \tag{3.3.22}$$

The fact that $F(v'^*, v)$ in (3.3.19) is an analytic function of both complex variables has a very important consequence. According to a theorem for functions of several complex variables[12], if a function $F(z, w)$, analytic in both variables in a neighborhood of the origin, vanishes in the subdomain $w = z^*$, it vanishes identically. It follows that, in (3.3.19), $F(v'^*, v) = 0$ for $v' = v$ implies $F(v'^*, v) \equiv 0$. Taking $F = F_1 - F_2$, where F_1 and F_2 are two operators having the same diagonal matrix elements, we conclude that *an operator F is uniquely determined by its diagonal matrix elements* $\langle v| F |v \rangle$ *in the coherent-state basis* (and even by their values in a neighborhood of the origin). It is natural to inquire, therefore, whether one can find a representation for F in terms of its diagonal matrix elements alone.

3.4 THE DIAGONAL REPRESENTATION

(a) Introduction

The simplest example of a diagonal coherent-state representation for an operator is the density operator for a pure coherent state

$$\varrho = |v_0\rangle \langle v_0|. \tag{3.4.1}$$

Another example is provided by the expansion (3.2.31) of the identity operator.

Let us now investigate under what conditions it is possible to represent a more general density operator by means of a *diagonal representation*

$$\varrho = \int \varphi(v) |v\rangle \langle v| \, d^2v, \tag{3.4.2}$$

where $\varphi(v)$, which must be real because of the hermiticity of ϱ, will be called the *weight function*. The density operator always has the non-diagonal representation (3.3.21),

$$\varrho = \frac{1}{\pi^2} \int d^2v' \int d^2v \exp\left[-\tfrac{1}{2}(|v|^2 + |v'|^2)\right] R(v'^*, v) |v'\rangle \langle v|. \tag{3.4.3}$$

However, due to the overcompleteness of the set of coherent states, we can find alternative representations, and a representation of the form (3.4.2), besides being simpler, would have many other advantages.

To give an example, suppose we want to compute the statistical average of the operator $(a^+)^m a^n$ in the state (3.4.2). According to (2.1.16) and (3.1.10), we have

$$\langle (a^+)^m a^n \rangle = \text{Tr}\left[\varrho(a^+)^m a^n\right] = \int d^2v \, \varphi(v) \langle v| (a^+)^m a^n |v\rangle$$

$$= \int \varphi(v) (v^*)^m v^n \, d^2v, \tag{3.4.4}$$

i.e., the evaluation of an operator trace is reduced to an ordinary (c-number) integration (a similar result would hold for (3.4.3), but in terms of a double integral). More generally, if $:\mathcal{O}(a^+, a):$ denotes any *normally-ordered operator*, defined by a power-series expansion

$$:\mathcal{O}(a^+, a): = \sum_{m,n} c_{m,n}(a^+)^m a^n, \qquad (3.4.5)$$

in each term of which the creation operators are always to the left of the annihilation operators, we have

$$\langle :\mathcal{O}(a^+, a): \rangle = \text{Tr}\,(\varrho:\mathcal{O}:) = \int \varphi(v)\,\mathcal{O}\,(v^*, v)\,d^2v, \qquad (3.4.6)$$

where $\mathcal{O}(v^*, v)$ is the *function* obtained by substituting $a \to v$, $a^+ \to v^*$ in (3.4.5). Since the coherence functions (2.3.1) are statistical averages of this form, it is obviously a great advantage to be able to employ a relation of the form (3.4.6).

To find out whether a diagonal representation exists, let us assume that (3.4.2) is valid and let us try to evaluate the weight function $\varphi(v)$ in terms of the diagonal matrix elements of ϱ (which, according to § 3.3, determine ϱ uniquely):

$$\langle v| \varrho |v\rangle = \int \varphi(v')\,|\langle v' | v\rangle|^2\,d^2v'$$

$$= \int \exp\,(-|v - v'|^2)\,\varphi(v')\,d^2v'. \qquad (3.4.7)$$

In order to solve this integral equation for the weight function $\varphi(v)$, we note that (3.4.7) looks like a convolution product, which suggests applying the Fourier transform. For this purpose, let us switch to real variables, with the help of (3.2.14), where we take $\omega = 1$ for simplicity:

$$v = (q + ip)/\sqrt{2}, \quad |v\rangle = |q, p\rangle. \qquad (3.4.8)$$

With this substitution, (3.4.7) becomes

$$\langle q, p| \varrho |q, p\rangle = \tfrac{1}{2} \int \exp\,\{-\tfrac{1}{2}[(q - q')^2 + (p - p')^2]\}\,\varphi(q', p')\,dq'\,dp'. \qquad (3.4.9)$$

Let us define the Fourier transforms

$$\tilde{\varrho}(x, k) = \frac{1}{2\pi} \int e^{i(xp-kq)}\,\langle q, p| \varrho |q, p\rangle\,dq\,dp, \qquad (3{:}4.10)$$

$$\tilde{\varphi}(x, k) = \frac{1}{2\pi} \int e^{i(xp-kq)}\,\varphi(q, p)\,dq\,dp, \qquad (3.4.11)$$

and let us employ the well-known result

$$\frac{1}{2\pi} \int e^{i(kq-xp)} \exp\,[-\tfrac{1}{2}(q^2 + p^2)]\,dq\,dp = \exp\,[-\tfrac{1}{2}(x^2 + k^2)]. \qquad (3.4.12)$$

Applying to (3.4.9) the convolution theorem for Fourier transforms, we get

$$\tilde{\varrho}(x, k) = \pi \exp\left[-\tfrac{1}{2}(x^2 + k^2)\right] \tilde{\varphi}(x, k),$$

and, solving with respect to $\tilde{\varphi}$, we get

$$\tilde{\varphi}(x, k) = \frac{1}{\pi} \exp\left[\tfrac{1}{2}(x^2 + k^2)\right] \tilde{\varrho}(x, k). \tag{3.4.13}$$

The weight function $\varphi(q, p)$ is then to be obtained by taking the inverse Fourier transform of (3.4.13). Conversely, if φ is defined in this way, it satisfies the integral equation (3.4.7), which, in view of the unique characterization of ϱ by its diagonal matrix elements, implies the validity of the diagonal representation (3.4.2).

At first sight, this seems to provide a complete solution of the problem. However, in order to justify the above formal manipulations, we have to investigate whether (and in what sense) the inverse Fourier transform of (3.4.13) is convergent.

(b) The optical "equivalence" theorem

It is already clear from the example (3.4.1) that we cannot confine our discussion of convergence to ordinary functions. In fact, in this case, with $v_0 = (q_0 + ip_0)/\sqrt{2}$,

$$\varphi(v) = \delta(v - v_0) = 2\delta(q - q_0)\,\delta(p - p_0) \tag{3.4.14}$$

is not a function, but rather a *distribution*. For the remainder of this section we assume that the reader is familiar with the elements of the theory of distributions[13]. Those who are less interested in the mathematical aspects may proceed directly to the formulation of the theorem at the end of this section.

It follows from (3.4.3) and (3.2.27) that

$$\mathrm{Tr}\,\varrho = \frac{1}{\pi^2} \int d^2v \exp\left(-|v|^2\right) \int \exp\left(v^*v' - |v'|^2\right) R(v'^*, v)\, d^2v'$$

$$= \frac{1}{\pi} \int \exp\left(-|v|^2\right) R(v^*, v)\, d^2v = \frac{1}{\pi} \int \langle v|\,\varrho\,|v\rangle\, d^2v$$

$$= \frac{1}{2\pi} \int \langle q, p|\,\varrho\,|q, p\rangle\, dq\, dp = 1, \tag{3.4.15}$$

where we have also employed (3.3.10) and (3.3.19). According to (3.4.10), this implies

$$|\tilde{\varrho}(x, k)| \leq 1, \tag{3.4.16}$$

so that (3.4.13) yields

$$|\tilde{\varphi}(x, k)| \leqq \frac{1}{\pi} \exp\left[\tfrac{1}{2}(x^2 + k^2)\right].$$ (3.4.17)

One can show, by means of specific examples[14], that this type of exponential of a quadratic behavior can actually be attained (although most cases of physical interest are much better behaved, as will be seen later). Can one define the Fourier transform of a function that may have such a rapid increase at infinity?

Since the right-hand side of (3.4.17) grows faster than any polynomial at infinity, the Fourier transform in question need not be a tempered distribution; it need not even be a Schwartz distribution, i.e., a distribution on the space $\mathscr{D}$ of infinitely differentiable test functions with compact support. However, it can be defined[15] as an *ultradistribution* or *analytic functional*[16]. For example,

$$e^{x^2} = \sum_{n=0}^{\infty} \frac{x^{2n}}{n!}$$

may be regarded as a Schwartz distribution, belonging to the Schwartz space $\mathscr{D}'$. Its Fourier transform

$$\mathscr{F}(e^{x^2}) = \sum_{n=0}^{\infty} \frac{(-1)^n}{n!} \delta^{(2n)}(k)$$ (3.4.18)

belongs to the space $\mathscr{Z}'$ of ultradistributions. It is an analytic functional on the space $\mathscr{Z}$; the test functions in $\mathscr{Z}$ are Fourier transforms of those in $\mathscr{D}$: they are (by the Paley-Wiener theorem) entire functions of exponential type which, along the real axis, decrease at infinity faster than any inverse power. The Fourier transformation is a one-to-one mapping between $\mathscr{D}$ and $\mathscr{Z}$, $\mathscr{D}'$ and $\mathscr{Z}'$.

Thus, if we interpret the weight function $\varphi(v)$ in (3.4.2) in the sense of ultradistributions, it is always given by the inverse Fourier transform of (3.4.13). The possibility of representing every density operator in diagonal form was asserted by Sudarshan[17], who gave a formal expression for the weight function in terms of a series of derivatives of δ of arbitrarily high orders [cf. (3.4.18)]; for the special case of thermal light, the diagonal representation was formulated by Glauber[18]. The mathematical convergence problems associated with the diagonal representation have given rise to a lengthy controversy in the literature.

Any representation of the density operator is ultimately to be applied to the evaluation of statistical averages, as indicated by (2.1.16). How is $\mathrm{Tr}\,(\varrho\mathscr{O})$ to be computed when the weight function in (3.4.2) is defined in terms of ultradistributions? As is well known, the more singular a distribution is, the more regular the corresponding class of test functions has to be. Ultradistributions can be so singular that the associated test function space $\mathscr{Z}$ is very restrictive, and this would lead to very severe restrictions on the class of operators for which $\text{Tr}\,(\varrho\mathcal{O})$ can be evaluated directly from (3.4.2) in this sense[19].

Nevertheless, one can take a different approach[20] to the problem, in which distributions are defined in terms of limits of sequences, and the aim from the outset is to define $\text{Tr}\,(\varrho\mathcal{O})$ in the same manner, i.e.,

$$\langle\mathcal{O}\rangle = \text{Tr}\,(\varrho\mathcal{O}) = \lim_{N\to\infty} \text{Tr}\,(\varrho_N\mathcal{O}), \qquad (3.4.19)$$

where $\{\varrho_N\}$ is a sequence of density operators converging to ϱ.

We may restrict ourselves to *bounded* operators[21]. In fact, given a hermitian operator $\mathcal{O}$, we may associate with it the unitary operator $e^{ix\mathcal{O}}$ and the *characteristic function*

$$C_\varrho(x) = \text{Tr}\,(\varrho e^{ix\mathcal{O}})$$

then contains the statistical information about the distribution of observable values associated with $\mathcal{O}$ in the state described by ϱ, as is well known from probability theory (for instance, expectation values of moments of $\mathcal{O}$ may be obtained from the derivatives of C_ϱ with respect to x taken at the origin). Thus, we may confine ourselves to bounded operators and, without restriction of generality, assume that

$$\|\mathcal{O}\| \leqq 1. \qquad (3.4.20)$$

A density operator is an example of a *trace-class operator*. Operators of this class have several important properties[22]. They form a linear space, so that, e.g., the difference $\varDelta_N = \varrho - \varrho_N$ in (3.4.19) is a trace-class operator. Every trace-class operator T can be put in the "polar form" $T = VA$ (analogous to the polar form $z = e^{i\theta}|z|$ for a complex number), where A, the analogue of the "modulus" of T (i.e., $(T^+T)^{\frac{1}{2}}$), is a positive hermitian operator with finite trace, and V is isometric, i.e., such that $V^+V = 1$. The operator A has properties similar to those of ϱ, and as a consequence can always be expanded in a form similar to (2.1.15), where $|\psi_n\rangle$ is a complete orthonormal set and the coefficients are non-negative. The operator V maps an orthonormal set $\{|\psi_n\rangle\}$ into another orthonormal set $\{|\varphi_n\rangle\}$. Consequently, one finds that every trace-class operator has the *canonical decomposition*

$$T = \sum_{n=1}^{\infty} \tau_n |\psi_n\rangle\langle\varphi_n|, \qquad (3.4.21)$$

where $\{|\psi_n\rangle\}$, $\{|\varphi_n\rangle\}$ are two complete orthonormal sets, and

$$\tau_n \geqq 0, \quad |\text{Tr}\,T| \leqq \sum_{n=1}^{\infty} \tau_n = \|T\|_1 < \infty. \qquad (3.4.22)$$

The norm $\|T\|_1$ is called *trace-class norm*.

Applying this result to $\Delta_N = \varrho - \varrho_N$ in (3.4.19), we find that

$$|\text{Tr } [(\varrho - \varrho_N) \, \mathcal{O}]| \leq \sum_{n=1}^{\infty} \tau_n \, |\langle \varphi_n| \, \mathcal{O} \, |\psi_n\rangle| \leq \|\mathcal{O}\| \sum_{n=1}^{\infty} \tau_n$$

$$\leq \|\varrho - \varrho_N\|_1, \tag{3.4.23}$$

with the help of (3.4.20) and (3.4.22). Consequently, if ϱ_N converges to ϱ in the trace-class norm, i.e., if

$$\lim_{N \to \infty} \|\varrho - \varrho_N\|_1 = 0, \tag{3.4.24}$$

the expectation values in (3.4.19) converge *uniformly* (i.e., independently of $\mathcal{O}$) to the desired statistical average.

Every trace-class operator can be written in the form $T = AB$, where A and B are *Hilbert-Schmidt operators*, i.e., operators with finite *Hilbert-Schmidt norm*

$$\|A\|_2 = [\text{Tr } (A^+A)]^{\frac{1}{2}}. \tag{3.4.25}$$

Furthermore, we have an inequality of the Schwarz type,

$$\|T\|_1 = \|AB\|_1 \leq \|A\|_2 \|B\|_2. \tag{3.4.26}$$

Since ϱ is a positive operator, it can be written in the form $\varrho = B^+B$, where $\|B\|_2 = 1$. We can then choose an approximating sequence of the form $\varrho_N = B_N^+ B_N$, with $\text{Tr } \varrho_N = 1$, so that each ϱ_N is a true density operator. By (3.4.26),

$$\|\varrho - \varrho_N\|_1 = \|B^+B - B_N^+ B_N\|_1 = \|B^+(B - B_N) + (B^+ - B_N^+) B_N\|_1$$

$$\leq \|B^+(B - B_N)\|_1 + \|(B^+ - B_N^+) B_N\|_1 \leq 2\|B - B_N\|_2. \tag{3.4.27}$$

Comparing this with (3.4.23), we see that it is sufficient to construct a sequence of operators B_N converging to B in Hilbert-Schmidt norm.

Now let us go back to (3.4.10) and invert it:

$$\langle q, p| \, \varrho \, |q, p\rangle = \frac{1}{2\pi} \int \tilde{\varrho}(x,k) e^{i(kq-xp)} \, dx \, dk. \tag{3.4.28}$$

Using (3.4.8) and introducing

$$z = (x + ik)/\sqrt{2}, \quad |z\rangle = |x, k\rangle, \tag{3.4.29}$$

we have, with the help of (3.2.6),

$$e^{i(kq-xp)} = e^{(zv^* - z^*v)} = \langle v| \, e^{za^+} e^{-z^*a} \, |v\rangle$$

$$= e^{\frac{1}{2}|z|^2} \langle v| \, U(z) \, |v\rangle = \exp [\tfrac{1}{4}(x^2 + k^2)] \langle q, p| \, U(x, k) \, |q, p\rangle, \tag{3.4.30}$$

where $U(x, k)$ is defined by (3.2.16). Substituting in (3.4.28) and taking into account that an operator is uniquely characterized by its diagonal matrix elements in the coherent-state basis, we get the operator equation

$$\varrho = \frac{1}{2\pi} \int r(x, k)\, U(x, k)\, dx\, dk, \tag{3.4.31}$$

where

$$r(x, k) = \exp\left[\tfrac{1}{4}(x^2 + k^2)\right] \tilde{\varrho}(x, k). \tag{3.4.32}$$

A representation of this form (Weyl representation[23]) is valid for any Hilbert-Schmidt operator, and in particular for B, in the decomposition $\varrho = B^+B$:

$$B = \frac{1}{2\pi} \int b(x, k)\, U(x, k)\, dx\, dk, \tag{3.4.33}$$

$$b(x, k) = \exp\left[\tfrac{1}{4}(x^2 + k^2)\right] \frac{1}{2\pi} \int e^{i(xp - kq)} \langle q, p|\, B\, |q, p \rangle\, dq\, dp. \tag{3.4.34}$$

It follows that, taking into account (3.2.2),

$$\varrho = B^+B = \frac{1}{4\pi^2} \int dx\, dk \int dx'\, dk'\, b^*(-x, -k)\, b(x', k')\, U(x', k')\, U(x, k)$$

$$= \frac{1}{4\pi^2} \int dx\, dk \int dx'\, dk'\, b^*(-x, -k)\, b(x', k') \exp\left[\frac{i}{2}(kx' - xk')\right]$$

$$\times U(x + x', k + k') = \frac{1}{2\pi} \int r(x, k)\, U(x, k)\, dx\, dk, \tag{3.4.35}$$

where

$$r(x, k) = \frac{1}{2\pi} \int b^*(x' - x, k' - k)\, b(x', k')$$

$$\times \exp\left[\frac{i}{2}(kx' - xk')\right] dx'\, dk', \tag{3.4.36}$$

and we have made use of (3.2.4), together with the commutation rule $[Q, P] = i$.

It follows from (3.4.15) and (3.4.28) that

$$\mathrm{Tr}\, \varrho = \frac{1}{4\pi^2} \int dx\, dk\, \tilde{\varrho}(x, k) \int dq\, dp\, e^{i(kq - xp)} = \tilde{\varrho}(0, 0), \tag{3.4.37}$$

so that, by (3.4.32) and (3.4.36),

$$1 = \mathrm{Tr}\, \varrho = (\|B\|_2)^2 = r(0, 0) = \frac{1}{2\pi} \int |b(x, k)|^2\, dx\, dk. \tag{3.4.38}$$

Thus, in the representation (3.4.33), the Hilbert-Schmidt norm of B is proportional to the L^2 norm of the weight function $b(x, k)$. The Fourier transform is always defined for such functions (this is the reason why (3.4.33) was restricted to Hilbert-Schmidt operators). It follows that (3.4.24) will be satisfied if we construct a sequence $\{b_N\}$ such that $b_N(x, k) \rightarrow b(x, k)$ in the L^2 norm.

According to a well-known result[24], however, the L^2-function $b(x, k)$ can always be approximated in the L^2 norm arbitrarily closely by a sequence of test functions $b_N(x, k)$ belonging to the space $\mathscr{D}$, i.e., infinitely differentiable and vanishing outside of a finite region. If we define $r_N(x, k)$ by (3.4.36) with b replaced by b_N, the function r_N also belongs to $\mathscr{D}$, because (3.4.36) is a kind of convolution product (it is known as a "twisted convolution product"), and the same is true for

$$\tilde{\varphi}_N(x, k) = \frac{1}{\pi} \exp \left[\tfrac{1}{4}(x^2 + k^2)\right] r_N(x, k)$$

$$= \frac{1}{\pi} \exp \left[\tfrac{1}{2}(x^2 + k^2)\right] \tilde{\varrho}_N(x, k), \tag{3.4.39}$$

which, according to (3.4.32) and (3.4.13), represents the Nth order approximation to the Fourier transform of the weight function $\varphi(q, p)$ of the diagonal representation. The main point is that the "dangerous" exponential factor has been rendered harmless in (3.4.39) by the fact that $\tilde{\varrho}_N(x, k)$ vanishes outside of a finite domain in the (x, k) plane.

We can now take the inverse Fourier transform, and, since $\mathscr{D}$ is contained in the space $\mathscr{S}$ of "rapidly decreasing" test functions (infinitely differentiable test functions that, together with all their derivatives, vanish at infinity faster than any inverse power), the inverse Fourier transform $\varphi_N(q, p)$ likewise belongs to $\mathscr{S}$.

We have thus finally arrived at the following result:

Optical "equivalence" theorem: For any density operator ϱ, and any bounded operator $\mathcal{O}$, $\|\mathcal{O}\| \leq 1$, there exists a sequence $\{\varrho_N\}$ of density operators with diagonal representations,

$$\varrho_N = \tfrac{1}{2} \int_{-\infty}^{\infty} \int_{-\infty}^{\infty} \varphi_N(q, p) |q, p\rangle \langle q, p| \, dq \, dp, \tag{3.4.40}$$

such that

$$\langle \mathcal{O} \rangle = \text{Tr}(\varrho \mathcal{O}) = \lim_{N \to \infty} \text{Tr}(\varrho_N \mathcal{O})$$

$$= \tfrac{1}{2} \lim_{N \to \infty} \int_{-\infty}^{\infty} \int_{-\infty}^{\infty} \varphi_N(q, p) \langle q, p| \mathcal{O} |q, p\rangle \, dq \, dp. \tag{3.4.41}$$

This sequence allows one to approximate $\langle \mathcal{O} \rangle$ uniformly (i.e., independently of $\mathcal{O}$), with arbitrary accuracy, by a diagonal representation having a real weight function $\varphi_N(q, p)$ which is infinitely differentiable and decreases faster than any inverse power of $(q^2 + p^2)^{\frac{1}{2}}$ as $(q^2 + p^2)^{\frac{1}{2}} \to \infty$.

In this way, the weight function $\varphi(v)$ in (3.4.2) is defined by the limit of a sequence $\{\varphi_N\}$ where each φ_N is an extremely well-behaved function. As will be seen below, $\varphi(v)$ itself is a well-behaved function in most of the practically important cases that have been considered so far. The reason for the name "optical equivalence theorem" will be discussed in § 3.4(d).

(c) The characteristic functions

The optical "equivalence" theorem, in the above form, cannot be directly applied to the evaluation of statistical averages like that in (3.4.4) (unless we know beforehand that $\varphi(v)$ is sufficiently well-behaved), because $(a^+)^m a^n$ is not a bounded operator. Nevertheless, we can always circumvent this problem, as indicated in § 3.4(b), by taking $\mathcal{O}$ to be the unitary operator $U(z) = U(x, k)$ defined by (3.2.6), (3.2.16), (3.4.29). We then get the *characteristic function*

$$C(z) = \text{Tr}\,[\varrho U(z)] = \text{Tr}\,(\varrho\, e^{za^+ - z^*a}). \tag{3.4.42}$$

With the help of (3.2.4), we find

$$C(z) = e^{-\frac{1}{2}|z|^2} C_N(z) = e^{\frac{1}{2}|z|^2} C_A(z), \tag{3.4.43}$$

where

$$C_N(z) = \text{Tr}\,(\varrho\, e^{za^+} e^{-z^*a}) \tag{3.4.44}$$

is the *normally-ordered characteristic function*, and

$$C_A(z) = \text{Tr}\,(\varrho\, e^{-z^*a} e^{za^+}) \tag{3.4.45}$$

is the *antinormally-ordered characteristic function*.

Expanding $C_N(z)$ in a power series, we find

$$C_N(z) = \sum_{n=0}^{\infty} \sum_{m=0}^{\infty} \frac{\tilde{G}^{(n,m)}}{n!\,m!} z^n (-z^*)^m, \tag{3.4.46}$$

where

$$\tilde{G}^{(n,m)} = \text{Tr}\,[\varrho(a^+)^n a^m]$$

$$= \left[\frac{\partial^n}{\partial z^n} \frac{\partial^m}{\partial(-z^*)^m} C_N \right]_{z=0} \tag{3.4.47}$$

is the function we want to evaluate.

Thus, if we evaluate $C(z)$, we can determine $C_N(z)$ from (3.4.43), and then $\tilde{G}^{(n,m)}$ follows from (3.4.47). On the other hand, since $U(z)$ is unitary, we can employ the optical "equivalence" theorem to evaluate $C(z)$ with the help of the diagonal representation (3.4.2):

$$C(z) = \int \varphi(v) \langle v| U(z) |v\rangle \, d^2v = e^{-\frac{1}{2}|z|^2} \int \varphi(v) \, e^{zv^* - z^*v} d^2v, \quad (3.4.48)$$

where we have employed (3.4.30) and, in general, the integral is to be interpreted symbolically, in the sense of (3.4.41).

Comparing (3.4.48) with (3.4.43), we see that

$$C_N(z) = \int \varphi(v) \, e^{zv^* - z^*v} \, d^2v. \qquad (3.4.49)$$

According to (3.4.30), this integral, written in terms of (x, k), (q, p), is nothing but a two-dimensional Fourier transform of $\varphi(p, q)$, i.e., by comparison with (3.4.11),

$$C_N(z) = C_N(x, k) = \tilde{\varphi}(-x, -k). \qquad (3.4.50)$$

The inverse Fourier transform is given by

$$\varphi(v) = \frac{1}{\pi^2} \int e^{vz^* - v^*z} C_N(z) \, d^2z. \qquad (3.4.51)$$

Thus, *the weight function of the diagonal representation and the normally-ordered characteristic function are Fourier transforms of each other* (again, this may have to be understood in a symbolic way).

The antinormally-ordered characteristic function (3.4.45) may be employed to evaluate statistical averages of antinormally-ordered operators, in a manner similar to (3.4.47). With the help of the resolution of unity (3.2.31), we find

$$C_A(z) = \frac{1}{\pi} \text{Tr} \int \{\varrho e^{-z^*a} |v\rangle \langle v| e^{za^+}\} d^2v$$

$$= \frac{1}{\pi} \int \langle v| \varrho |v\rangle \, e^{zv^* - z^*v} \, d^2v. \qquad (3.4.52)$$

Taking into account (3.4.30) and (3.4.10), we conclude that

$$C_A(z) = C_A(x, k) = \tilde{\varrho}(-x, -k). \qquad (3.4.53)$$

According to (3.4.43),

$$C_N(z) = e^{|z|^2} C_A(z). \qquad (3.4.54)$$

We see from (3.4.50) and (3.4.53), that (3.4.54) is nothing but a restatement of (3.4.13).

On the other hand, (3.4.32), (3.4.43), and (3.4.53) yield

$$C(z) = C(x, k) = r(-x, -k), \qquad (3.4.55)$$

so that (3.4.31) may be rewritten as

$$\varrho = \frac{1}{\pi} \int C(z)\, U(-z)\, d^2z. \tag{3.4.56}$$

The relations (3.4.42) and (3.4.56) may be regarded as a kind of *Fourier transform pair for operators*.

Since $C(z)$ is the expectation value (3.4.42) of a unitary operator $U(z)$, with $U(0) = 1$, it must satisfy [cf. (3.4.55) and (3.4.38)]

$$|C(z)| \leqq 1, \qquad C(0) = \mathrm{Tr}\, \varrho = 1. \tag{3.4.57}$$

It then follows from (3.4.43) that

$$|C_A(z)| \leqq e^{-\frac{1}{2}|z|^2}, \qquad |C_N(z)| \leqq e^{\frac{1}{2}|z|^2}. \tag{3.4.58}$$

The possibility of this exponential growth for $C_N(z)$ indicates once more that (3.4.51) may have to be interpreted symbolically, as discussed in the previous section.

It is clear from the above discussion that the knowledge of a characteristic function is essentially equivalent to that of the density operator, i.e., to a complete specification of the state of the system.

(d) Extension to several modes

So far we have confined our discussion to a single mode of the radiation field. In general, however, any number of modes may be excited: the field has an infinite number of degrees of freedom.

In order to extend the definition of coherent states to the general case, we can again consider their expansion in the basis of Fock states,

$$|n_1, n_2, \ldots\rangle = |\{n_k\}\rangle, \tag{3.4.59}$$

where n_k is the occupation number of the kth mode [cf. (2.4.22)]. We restrict the sequences $\{n_k\}$ by the condition

$$\sum_{k=1}^{\infty} n_k < \infty, \tag{3.4.60}$$

so that, for any basis vector, only a finite number of n_k are non-vanishing. The basis vectors are, therefore, countable (separable Hilbert space).

The coherent states are now defined by [cf. (3.1.9), (3.1.24)],

$$|v_1, v_2, \ldots\rangle = |\{v_k\}\rangle = \sum_{\{n_k\}=0}^{\infty} \left[\prod_{k'=1}^{\infty} \exp(-\tfrac{1}{2}|v_{k'}|^2) \frac{(v_{k'})^{n_{k'}}}{\sqrt{n_{k'}!}} \right] |\{n_k\}\rangle, \tag{3.4.61}$$

where we assume that

$$U = \sum_{k'=1}^{\infty} |v_{k'}|^2 < \infty. \tag{3.4.62}$$

According to (3.1.26), U should represent the average total number of photons, so that (3.4.62) amounts to assuming that this number is finite[25]. The sequences $\{v_k\}$ may thus also be regarded as elements of a Hilbert space. In particular, (3.4.62) implies

$$\lim_{k' \to \infty} (a_{k'} \mid \{v_k\}\rangle) = \lim_{k' \to \infty} (v_{k'} \mid \{v_k\}\rangle) = 0, \qquad (3.4.63)$$

so that, for sufficiently large k', the coherent state for the corresponding mode must approach the ground (vacuum) state.

The extension of the diagonal representation (3.4.2) to the general case may be written symbolically as

$$\varrho = \int \varphi(\{v_k\}) \mid \{v_k\}\rangle \langle\{v_k\}\mid d^2\{v_k\}, \qquad (3.4.64)$$

where

$$d^2\{v_k\} = \prod_k d^2 v_k \qquad (3.4.65)$$

and the integral over infinitely many variables is to be interpreted, in agreement with (3.4.63), as the limit of an integral over the first M modes (the remaining ones being in the ground state) as $M \to \infty$. For each such finite integral, we can apply the optical "equivalence" theorem [cf. (3.4.40)], so that (3.4.64) involves[26] a double limit ($M \to \infty$, $N \to \infty$).

We can now go back to (2.4.22) and conclude that

$$E^{(+)}(x) \mid \{v_k\}\rangle = \sum_k f_k(x) a_k \mid \{v_k\}\rangle$$

$$= \sum_k v_k f_k(x) \mid \{v_k\}\rangle = V(x) \mid \{v_k\}\rangle, \qquad (3.4.66)$$

where

$$f_k(x) = i\sqrt{\frac{\omega_k}{2}} u_k(\mathbf{r}) \exp(-i\omega_k t) \qquad (3.4.67)$$

and $V(x)$ is defined by (3.1.6). In this way we associate an analytic signal $V(x)$ with the annihilation part of the electric field operator $E^{(+)}(x)$ for every coherent state $\mid \{v_k\}\rangle$.

We can now apply this result, together with (2.3.1) and the diagonal representation (3.4.64), to conclude that

$$G^{(n,m)}(x_1, \ldots, x_n; x_{n+1}, \ldots, x_{n+m})$$

$$= \mathrm{Tr}\,[\varrho E^{(-)}(x_1) \ldots E^{(-)}(x_n) E^{(+)}(x_{n+1}) \ldots E^{(+)}(x_{n+m})]$$

$$= \langle V^*(x_1) \ldots V^*(x_n) V(x_{n+1}) \ldots V(x_{n+m})\rangle$$

$$= \Gamma^{(n,m)}(x_1, \ldots, x_n; x_{n+1}, \ldots, x_{n+m}), \qquad (3.4.68)$$

where we have employed (3.4.6), and

$$\langle V^*(x_1) \dots V(x_{n+m}) \rangle = \int \varphi(\{v_k\}) \, V^*(x_1) \dots V(x_{n+m}) \, d^2\{v_k\}, \qquad (3.4.69)$$

each V being expressed in terms of the $\{v_k\}$ by (3.1.6). The notation $\Gamma^{(n,m)}$ in (3.4.68) is a natural extension of (1.3.45): we may regard (3.4.69) as a kind of "classical" statistical average, with the weight function $\varphi(\{v_k\})$ playing the role of a "probability distribution" in the phase space of the variables $\{v_k\}$.

It has been assumed in (3.4.68) that the diagonal representation is sufficiently regular so that we can apply the "equivalence" theorem to the unbounded operator that appears in this expression. This is usually true in practice; however, even in cases where it is not, we can still apply the theorem to the normally-ordered characteristic function, as we have seen in the preceding section, and [cf. (3.4.47)] this is sufficient to determine the quantities

$$\tilde{G}^{(n,m)}(k_1, \dots, k_n; k_{n+1}, \dots, k_{n+m}) = \text{Tr} \, (\varrho a_{k_1}^+ \dots a_{k_n}^+ \, a_{k_{n+1}} \dots a_{k_{n+m}}), \qquad (3.4.70)$$

which are Fourier transforms of the coherence functions (3.4.68).

Thus, the diagonal representation induces a complete *formal* equivalence between the expressions found in the quantum theory of coherence and those found in the classical theory, with the weight function φ playing the role of "probability distribution". This is the reason for the name given to the optical "equivalence" theorem. Let us now explain why "equivalence" has been placed within quotation marks.

(e) Discussion

There are several reasons why the "equivalence" between the classical and quantum theories of coherence expressed by the optical "equivalence" theorem must usually be understood in a purely formal sense, and one must be aware of the limitations of this analogy in order not to be misled into transgressing beyond its limits of validity.

(1) The physical content of the quantum theory is of course vastly different from that of classical theory. So far, we have discussed only the description of the coherence properties of a *given* ensemble of fields, without asking how this ensemble is determined. If quantum effects play a significant role in the generation of the field by a given source, the weight function φ in the diagonal representation must be obtained by solving a fully quantum-mechanical problem to find the density operator of the system. The results of this quantum-mechanical calculation can then be described in terms of the function φ, that looks like a classical "probability distribution". However, as will now be seen, φ cannot be interpreted, in general,

as a true probability distribution. To see this, it suffices to restrict the discussion to a single mode.

(2) In spite of the formal analogy between (3.4.2) and (2.1.15), we cannot interpret $\varphi(v)$ as the "probability of finding the system in the state $|v\rangle$" because the projectors $|v\rangle \langle v|$ are not orthogonal for different values of v. This is also clear from (3.2.15): one cannot speak of a *point* (p, q) in the "phase space" of the oscillator, because P and Q do not commute. As we have seen in § 3.2(d), one should rather interpret $|v\rangle$, in agreement with the uncertainty principle, as corresponding to a kind of unit cell in phase space. When two such cells do not overlap significantly, the corresponding states, by (3.2.28), are approximately orthogonal. Accordingly, if we have a "slowly-varying" weight function $\varphi(v)$, without appreciable variation within a unit cell, it may be approximately justified to interpret it as a probability density. This usually happens in the region $|v| \gg 1$ (large average number of photons), corresponding to fields that may be approximately described in classical terms.

(3) The weight function $\varphi(v)$ satisfies the normalization condition [cf. (3.4.2)],

$$\operatorname{Tr} \varrho = \int \varphi(v)\, d^2 v = 1. \tag{3.4.71}$$

However, in contrast with a true probability distribution, $\varphi(v)$ need not be positive semidefinite; it may also take on negative values [this can be shown explicitly by means of examples; cf. § 4.2(b)]. According to (3.4.7) and (2.1.27), the positive character of ϱ implies only

$$\langle v| \varrho |v\rangle = \int \exp\left(-|v - v'|^2\right) \varphi(v')\, d^2 v' \geqq 0. \tag{3.4.72}$$

This restricts the size of the regions of the v-plane over which φ can be negative (they cannot be much larger than the above-mentioned unit cells), but it does not exclude negative values. The right-hand side of (3.4.72), which is a kind of average of $\varphi(v)$ over a unit cell, is non-negative; thus, we see again that if $\varphi(v)$ is slowly-varying, the probability interpretation is justified.

(4) The diagonal representation is only one of the many possible ways in which one can represent density operators by means of weight functions $W(v)$ in such a way that $\operatorname{Tr}(\varrho\mathcal{O})$ can be evaluated as a "classical" average, with weight $W(v)$, of a *function* $\mathcal{O}(v)$ associated with suitably ordered operators $\mathcal{O}$. As shown by (3.4.6), the function $\varphi(v)$ is suitable for normally-ordered operators $\mathcal{O}$, and this is the reason for its utility in coherence theory. The rule of association between functions and operators depends on the prescription adopted for ordering the operators, which is obviously not unique.

5*

The earliest example of such a "generalized phase-space representation" was given by Wigner[27]; it corresponds to the well known "Weyl rule of association" between functions and operators, in which a product of operators is defined as the average of all possible orderings (e.g., aa^+a is replaced by $^1/_3(a^2a^+ + aa^+a + a^+a^2)$). The Wigner representation is closely related with the Weyl representations (3.4.33), (3.4.56) that we have already employed.

Similarly, one can show that the function $\langle v| \varrho |v\rangle$ corresponds to the weight function associated with antinormal ordering [an example is provided by (3.4.52)]. The weight functions associated wity such generalized phase-space representations are known as "quasiprobability distributions", because they cannot usually be interpreted as true probabilities. General discussions of such representations have recently been given[28,29].

(5) The weight function $\varphi(v)$ may be very singular, as we have seen in § 3.4(b), so that we have to define it as the limit of a sequence, as in (3.4.41). On the other hand, one can always use the nondiagonal representation (3.4.3), which also reduces the evaluation of the statistical average of a normally-ordered operator to a (double) integral over the v-plane, in which $R(v'^*, v)$ is very well behaved (it is an entire function of both variables).

In spite of the above limitations, the diagonal representation is an extremely valuable tool. In most applications, the weight function is found to be quite well behaved. The optical "equivalence" theorem, employed with due caution, furnishes valuable insights into the connection between the classical and quantum descriptions, and it allows us to make use of classical results, via the correspondence principle, in order to extend them into the quantum domain.

References

1. U. M. Titulaer and R. J. Glauber, *Phys. Rev.* **145**, 1041 (1966).
2. Cf., e.g., W. Heitler, *The Quantum Theory of Radiation*, 3rd ed., Oxford Univ. Press (1954), Chapter 1; W. Louisell, *Radiation and Noise in Quantum Electronics*, McGraw-Hill, New York (1964), Chapter 4.
3. See, however, P. Carruthers and M. M. Nieto, *Rev. Mod. Phys.* **40**, 411 (1968).
4. Cf., e.g., A. Messiah, *Quantum Mechanics*, vol. I, North-Holland Publishing Co., Amsterdam (1964), p. 442.
5. E. Schrödinger, *Naturwissenschaften* **14**, 664 (1926). Cf., also, E. M. Henley and W. Thirring, *Elementary Quantum Field Theory*, McGraw-Hill, New York (1962), Chapter 2.
6. R. J. Glauber, *Phys. Rev.* **131**, 2766 (1963).
7. E. H. Kennard, *Z. Physik* **44**, 326 (1927). Cf., also, W. Heisenberg, *The Physical Principles of the Quantum Theory*, Univ. of Chicago Press, Chicago (1930), p. 19.
8. Cf. P. Carruthers and M. M. Nieto, *Am. J. Phys.* **33**, 537 (1965).

9. R. J. Glauber, *Phys. Letters* **21**, 650 (1966); C. L. Mehta, P. Chand, E. C. G. Sudarshan, and R. Vedam, *Phys. Rev.* **157**, 1198 (1967).

10. An entire function $f(z)$ is of order ϱ if

$$f(z) = O[\exp(|z|^{\varrho + \varepsilon})] \quad \text{as} \quad |z| \to \infty$$

for every $\varepsilon > 0$ (no matter how small), but not for any $\varepsilon < 0$; cf., e.g., E. C. Titchmarsh, *The Theory of Functions*, 2nd ed., Oxford Univ. Press (1939), p. 248.

11. V. Bargmann, *Commun. Pure Appl. Math.* **14**, 187 (1961); cf. also I. E. Segal, *Illinois J. Math.* **6**, 500 (1962).

12. S. Bochner and W. T. Martin, *Several Complex Variables*, Princeton University Press (1948), p. 36. This is analogous to the theorem that an analytic function of a single complex variable cannot vanish over any real interval without vanishing identically.

13. Cf. L. Schwartz, *Mathematics for the Physical Sciences*, Addison-Wesley Publishing Co., Reading, Mass. (1966); I. M. Gel'fand and G. E. Shilov, *Generalized Functions*, vol. 1, Academic Press, New York (1965).

14. One such example is a pure state corresponding to a general minimum-uncertainty wave packet (3.2.19), with q_0 not related to ω by (3.1.20) [cf. *KS* (see Ch. 1, Ref. 4), p. 193].

15. M. M. Miller and E. A. Mishkin, *Phys. Rev.* **164**, 1610 (1967); M. M. Miller, *J. Math. Phys.* **9**, 1270 (1968). Cf., also, K. E. Cahill, *Phys. Rev.* **180**, 1244 (1969).

16. Cf. I. M. Gel'fand and G. E. Shilov, op. cit.; also, W. Güttinger, *Fortschr. der Phys.* **14**, 483 (1966).

17. E. C. G. Sudarshan, *Phys. Rev. Letters* **10**, 277 (1963).

18. R. J. Glauber, *Phys. Rev. Letters* **10**, 84 (1963).

19. Cf. K. E. Cahill, *Phys. Rev.* **180**, 1244 (1969).

20. J. R. Klauder, *Phys. Rev. Letters* **16**, 534 (1966); F. Rocca, *Compt. Rend.* **262**, A 547 (1966); cf., also, *KS*, p. 185.

21. An operator $\mathcal{O}$ is bounded if $\sup |\langle\varphi|\mathcal{O}|\psi\rangle| = \|\mathcal{O}\| < \infty$ for all normed vectors $|\varphi\rangle$, $|\psi\rangle$.

22. Cf. R. Schatten, *Norm Ideals of Completely Continuous Operators*, Springer-Verlag, Berlin (1960), and I. M. Gel'fand and N. Ya. Vilenkin, *Generalized Functions*, vol. 4, Academic Press, New York (1964).

23. H. Weyl, *The Theory of Groups and Quantum Mechanics*, Dover, New York (1931), p. 274; J. E. Moyal, *Proc. Cambridge Phil. Soc.* **45**, 99 (1949); K. E. Cahill and R. J. Glauber, *Phys. Rev.* **177**, 1857 (1969).

24. Cf., e.g., I. M. Gel'fand and G. E. Shilov, op. cit., Appendix 1.

25. The above assumptions have to be weakened when one wants to treat the problem of the infrared catastrophe in quantum electrodynamics (one has to deal with an infinite number of soft photons); cf. T. W. B. Kibble, *J. Math. Phys.* **9**, 315 (1968); *Phys. Rev.* **173**, 1527 (1968); **174**, 1882 (1968); **175**, 1624 (1968).

26. For further details, cf. *KS*, p. 195.

27. E. P. Wigner, *Phys. Rev.* **40**, 749 (1932). Cf., also, J. E. Moyal, *Proc. Cambridge Phil. Soc.* **45**, 99 (1949).

28. K. E. Cahill and R. J. Glauber, *Phys. Rev.* **177**, 1857; 1882 (1969).

29. E. Wolf and G. S. Agarwal, in *Polarisation, Matière et Rayonnement*, Presses Universitaires de France, Paris (1969), p. 541; also, *Phys. Rev. D* **2**, 2161, 2187, 2206 (1970).

CHAPTER 4

Applications

4.1 EXAMPLES OF LIGHT FIELDS

(a) Thermal light

As A FIRST example of application of the diagonal representation, let us consider the case of thermal light, i.e., light emitted by a source in thermal equilibrium at a temperature T. The corresponding density operator is given by (2.3.14), where H is the hamiltonian of the free radiation field, given by (neglecting zero-point energy)

$$H = \sum_k \omega_k a_k^+ a_k \tag{4.1.1}$$

in terms of the mode expansion (2.4.22). For simplicity, let us consider first the case of a single mode, taking

$$H = \omega a^\dagger a. \tag{4.1.2}$$

This is diagonal in the Fock basis $\{|n\rangle\}$, so that

$$\langle m| e^{-\beta H} |n\rangle = e^{-\beta n \omega} \delta_{m,n}, \tag{4.1.3}$$

$$\mathrm{Tr}\,(e^{-\beta H}) = \sum_{n=0}^{\infty} (e^{-\beta \omega})^n = (1 - e^{-\beta \omega})^{-1}, \tag{4.1.4}$$

and (2.3.14) becomes

$$\varrho = (1 - e^{-\beta \omega}) \sum_{n=0}^{\infty} (e^{-\beta \omega})^n |n\rangle \langle n|. \tag{4.1.5}$$

This leads to the well known Planck's law result

$$\langle n \rangle = \mathrm{Tr}\,(\varrho a^+ a) = \frac{1}{e^{\beta \omega} - 1} \tag{4.1.6}$$

in terms of which (4.1.5) may be rewritten as

$$\varrho = \sum_{m=0}^{\infty} p(m) |m\rangle \langle m|, \tag{4.1.7}$$

where $p(m)$, the probability of finding m photons in the field, is given by the *Bose distribution*

$$p(m) = \frac{1}{(1 + \langle n \rangle)} \left(\frac{\langle n \rangle}{1 + \langle n \rangle} \right)^m. \tag{4.1.8}$$

Let us now find the diagonal coherent-state representation for ϱ. This can be done directly by changing from (4.1.7) to the coherent basis, but let us follow, instead, the procedure indicated in § 3.4. We start by computing $\langle v | \varrho | v \rangle$, which, by (4.1.5) and (3.1.24), is given by

$$\langle v | \varrho | v \rangle = (1 - \zeta) \sum_{n=0}^{\infty} \zeta^n |\langle n | v \rangle|^2 = (1 - \zeta) e^{-|v|^2} \sum_{n=0}^{\infty} \frac{(\zeta |v|^2)^n}{n!}$$

$$= (1 - \zeta) \exp[-(1 - \zeta) |v|^2], \tag{4.1.9}$$

where we have introduced the abbreviation

$$\zeta = e^{-\beta \omega}. \tag{4.1.10}$$

It then follows from (3.4.8), (3.4.10) and (3.4.12), that

$$\tilde{\varrho}(x, k) = \exp\left[-\frac{(x^2 + k^2)}{2(1 - \zeta)} \right], \tag{4.1.11}$$

so that (3.4.13) becomes

$$\tilde{\varphi}(x, k) = \frac{1}{\pi} \exp\left[-\frac{1}{2} \frac{\zeta}{(1 - \zeta)} (x^2 + k^2) \right]$$

$$= \frac{1}{\pi} \exp\left[-\frac{\langle n \rangle (x^2 + k^2)}{2} \right]. \tag{4.1.12}$$

This is again a Gaussian, so that the inverse Fourier transform is a perfectly well-behaved function. By (3.4.11) and (3.4.12), we find

$$\varphi(v) = \frac{1}{\pi \langle n \rangle} \exp\left(-\frac{|v|^2}{\langle n \rangle} \right), \tag{4.1.13}$$

and the density operator is given by

$$\varrho = \frac{1}{\pi \langle n \rangle} \int \exp\left(-\frac{|v|^2}{\langle n \rangle} \right) |v\rangle \langle v| \, d^2v. \tag{4.1.14}$$

If all the modes are taken into account, we find [cf. (3.4.64)]

$$\varphi(\{v_k\}) = \prod_k \frac{1}{\pi \langle n_k \rangle} \exp\left(-\frac{|v_k|^2}{\langle n_k \rangle} \right). \tag{4.1.15}$$

Thus, the weight function is an ordinary, well-behaved function, and it is even positive definite. The fact that it depends only on the absolute values $|v_k|$ expresses the *stationary* character of the field. In fact, when we employ (4.1.15) to compute, e.g., $\langle a_k^+ a_{k'} \rangle$ [cf. (3.4.70)], the integrals over the phases of the v_k vanish unless $k = k'$, so that different frequencies are uncorrelated [cf. (1.3.32)].

The fact that we have found for the weight function associated with thermal light a product of Gaussians has a very simple physical interpretation. A thermal light source may be thought of as a collection of a very large number of independently emitting atoms. According to the Central Limit Theorem in the theory of stochastic processes, the resultant of a very large number of independent random variables tends to be characterized by a Gaussian probability distribution, so that (4.1.13) can be interpreted as a distribution associated with a kind of random walk in the v-plane. This interpretation also suggests that (4.1.15) has a much wider range of applicability: without the restriction that $\langle n_k \rangle$ is given by Planck's law, it should describe not only thermal light, but also light emitted by any kind of *chaotic source*, not necessarily in thermal equilibrium[1] (such as a gas discharge tube).

(b) Ideal laser light

Although we will discuss the laser in detail later on, it is useful at this point to introduce a highly idealized model of laser radiation. As will be seen later, the atoms of the active medium in a laser may be thought of as electric dipoles that oscillate coherently with the field, while the field in its turn results from their radiation by induced emission. The dipole moments of the radiating atoms give rise to an oscillating macroscopic density of polarization $\mathbf{P}$. As $\partial \mathbf{P}/\partial t$ is associated with a current density $\mathbf{j}_p$, we can also describe the field as that of a macroscopic oscillating current distribution (a sort of "optical antenna"). Due to nonlinear stabilization effects to be discussed later, the amplitude of this current is practically constant.

Thus, in this idealized model, the field may be regarded as that of a classical (c-number) predetermined current distribution $\mathbf{j}(\mathbf{r}, t)$. The problem of the radiation from such a distribution corresponds, for each mode, to that of a forced harmonic oscillator, so that we expect, according to § 3.2(c), that it will bring the field into a coherent state.

The interaction hamiltonian H_I in the Schrödinger equation (2.2.11) is now given by

$$H_I(t) = - \int \mathbf{j}(\mathbf{r}, t) \cdot \mathbf{A}(\mathbf{r}, t) \, d^3r. \tag{4.1.16}$$

Since $\mathbf{j}$ here is a c-number, the commutator $[H_I(t), H_I(t')]$ is a c-number. This allows us[2] to integrate (2.2.11) almost as if it were a numerical differential equation, to get

$$|t\rangle = \exp \left[i \int_{t_0}^{t} dt' \int \mathbf{j}(\mathbf{r}, t') \cdot \mathbf{A}(\mathbf{r}, t') \, d^3r + i\theta(t) \right] |0\rangle, \tag{4.1.17}$$

where $\theta(t)$ is a c-number phase term, arising from the commutator $[H_I(t'), H_I(t'')]$, via (3.2.4), that drops out when we evaluate the density operator, and we have assumed that, initially (at $t = t_0$), the field was in the vacuum state.

Substituting the mode expansion

$$\mathbf{A}(\mathbf{r}, t) = \sum_k [a_k \mathbf{v}_k(\mathbf{r}, t) + a_k^+ \mathbf{v}_k^*(\mathbf{r}, t)], \tag{4.1.18}$$

we find, with the help of (3.2.7),

$$e^{-i\theta(t)} |t\rangle = \prod_k U(v_k(t)) |0\rangle = \prod_k |v_k(t)\rangle, \tag{4.1.19}$$

where $U(v)$ is the unitary operator (3.2.6), and

$$v_k(t) = i \int_{t_o}^{t} dt' \int d^3r \, \mathbf{j}(\mathbf{r}, t') \cdot \mathbf{v}_k^*(\mathbf{r}, t'). \tag{4.1.20}$$

We see from (4.1.19) that the field produced by a classical current distribution indeed corresponds to a (time-dependent) coherent state.

In particular, if only one mode is excited, we find that the density operator for the field is given by (3.4.1), $\varrho = |v_0\rangle \langle v_0|$, which corresponds to a diagonal representation with the weight function (3.4.14), $\varphi(v) = \delta(v - v_0)$.

The above results are based on the assumption that both the amplitude and the phase of the current distribution are precisely known. In practice, however, while the amplitude is well determined by the stabilization effects referred to above, we do not have any phase information [cf. § 7.6(c)]. Thus, a more appropriate form of the density operator would be obtained by taking an average over all phases with uniform weight, i.e., for a single mode,

$$\varrho = \frac{1}{2\pi} \int_0^{2\pi} ||v_0| \, e^{i\varphi}\rangle \langle |v_0| \, e^{i\varphi} | \, d\varphi = \frac{1}{2\pi} \int_0^{2\pi} d\varphi \int_0^{\infty} |v| \, d|v|$$

$$\times \frac{1}{|v|} \delta(|v| - |v_0|) |v\rangle \langle v|, \tag{4.1.21}$$

corresponding to a diagonal representation with the weight function

$$\varphi(v) = \frac{1}{2\pi |v|} \delta(|v| - |v_0|), \tag{4.1.22}$$

which depends only on $|v|$, and therefore again represents a stationary field. Taking into account that the remaining modes are not excited, we get the weight function [cf. (3.4.64)]

$$\varphi(\{v_k\}) = \frac{1}{2\pi |v_k|} \delta(|v_k| - |v_{k,0}|) \prod_{k' \neq k} \frac{1}{2\pi |v_{k'}|} \delta(|v_{k'}|). \tag{4.1.23}$$

It is readily seen that such a field, just like (4.1.19), is still *fully coherent* (to all orders), i.e., (2.4.21) is still satisfied for all n. In fact, as we saw in

§ 2.4, the results are not affected by a constant phase factor in $V(x)$. This would not be true for the non-diagonal coherence functions (3.1.4); Glauber's definition of full coherence was deliberately chosen so as to be insensitive to the choice of an overall phase factor.

4.2 PHOTON COUNTING PROBABILITIES

(a) The counting distribution

We have already seen in (3.1.26) that the probability of finding n photons in a coherent state for a single mode is given by a Poisson distribution. This result can readily be extended to any number of modes. The probability $p(\{n_k\})$ of finding a given distribution $\{n_k\}$ for the numbers of photons in the various modes is the expectation value of the projection operator $|\{n_k\}\rangle \langle\{n_k\}|$ (which takes the value 1 for the given distribution and 0 for any other one), i.e.,

$$p(\{n_k\}) = \text{Tr} \left(\varrho \, |\{n_k\}\rangle \langle\{n_k\}|\right). \tag{4.2.1}$$

Substituting ϱ by the diagonal representation (3.4.64), we find

$$p(\{n_k\}) = \int \varphi(\{v_k\}) \, |\langle\{v_k\} \mid \{n_k\}\rangle|^2 \, d^2\{v_k\}$$

$$= \int \varphi(\{v_k\}) \prod_k \frac{|v_k|^{2n_k}}{n_k!} \exp\left(-|v_k|^2\right) d^2\{v_k\}, \tag{4.2.2}$$

which is an average over a product of Poisson distributions similar to (3.1.26).

Let $p(n)$ be the probability to find a total number n of photons in the field, distributed in any manner whatsoever over the various modes. Clearly,

$$p(n) = \sum_{\{n_k\},\, \sum_k n_k = n} p(\{n_k\}). \tag{4.2.3}$$

On the other hand, by the well known multinomial formula,

$$\left(\sum_k |v_k|^2\right)^n = \sum_{\{n_k\},\, \sum_k n_k = n} n! \prod_k \frac{|v_k|^{2n_k}}{n_k!}, \tag{4.2.4}$$

so that (4.2.2) and (4.2.3) yield [cf. (3.4.69)]

$$p(n) = \int \varphi(\{v_k\}) \frac{U^n}{n!} e^{-U} d^2\{v_k\} = \left\langle \frac{U^n}{n!} e^{-U} \right\rangle, \tag{4.2.5}$$

where, as in (3.4.62),

$$U = \sum_k |v_k|^2 \tag{4.2.6}$$

is proportional to the integral of the total intensity over the whole field (average total number of photons in the coherent state $|\{v_k\}\rangle$).

Thus, the probability distribution for the total number of photons is a kind of statistical average (with weight function φ) over Poisson distributions. This result was obtained by Mandel[3].

The average total number of photons is

$$\langle n \rangle = \sum_{n=1}^{\infty} n p(n) = \left\langle U e^{-U} \sum_{n=1}^{\infty} \frac{U^{n-1}}{(n-1)!} \right\rangle = \langle U \rangle. \qquad (4.2.7)$$

Similarly,

$$\langle n^2 \rangle = \sum_{n=1}^{\infty} n^2 p(n) = \left\langle U^2 e^{-U} \sum_{n=2}^{\infty} \frac{U^{n-2}}{(n-2)!} \right\rangle + \left\langle U e^{-U} \sum_{n=1}^{\infty} \frac{U^{n-1}}{(n-1)!} \right\rangle,$$

i.e.,

$$\langle n^2 \rangle = \langle U^2 \rangle + \langle U \rangle. \qquad (4.2.8)$$

If we define

$$\langle (\varDelta U)^2 \rangle = \langle U^2 \rangle - \langle U \rangle^2, \qquad (4.2.9)$$

it then follows that

$$\langle (\varDelta n)^2 \rangle = \langle n^2 \rangle - \langle n \rangle^2 = \langle n \rangle + \langle (\varDelta U)^2 \rangle. \qquad (4.2.10)$$

The first term, $\langle n \rangle$, is the fluctuation one would have if photons were independent classical particles. The second term, $\langle (\varDelta U)^2 \rangle$, is associated with the intensity fluctuations one would have in a classical wave field of intensity proportional to U. Both the corpuscular and the wave aspects contribute to the fluctuations. This is a generalization of a result due to Einstein[4] on energy fluctuations in blackbody radiation. We see that it holds not only for thermal radiation, but for any kind of light field.

(b) Examples

Let us apply (4.2.10) to some examples of light fields. For simplicity we confine our attention to a single mode.

(i) THERMAL LIGHT In this case, $p(n)$ is given by the Bose distribution (4.1.8). Taking into account the power series expansion

$$\sum_{n=1}^{\infty} n^2 x^n = \frac{x(1+x)}{(1-x)^3} \qquad (|x| < 1), \qquad (4.2.11)$$

we find

$$\langle n^2 \rangle = \frac{1}{1 + \langle n \rangle} \sum_{n=1}^{\infty} n^2 \left(\frac{\langle n \rangle}{1 + \langle n \rangle} \right)^n = \langle n \rangle + 2 \langle n \rangle^2, \qquad (4.2.12)$$

so that

$$\langle (\varDelta n)^2 \rangle = \langle n \rangle + \langle n \rangle^2. \qquad (4.2.13)$$

In this case, therefore, the second term of (4.2.10) is positive: the fluctuations are greater than they would be for independent particles. This is due to the effects of Bose-Einstein statistics.

(ii) IDEAL LASER LIGHT   In this case, regardless of whether one adopts
(3.4.1) or (4.1.21) for the density operator, the "intensity" (4.2.6) is constant, so that (4.2.5) becomes

$$p(n) = \frac{U_0^n}{n!} e^{-U_0},\qquad (4.2.14)$$

i.e., we have a Poisson distribution. Thus, $\langle (\varDelta U) \rangle^2 = 0$, and (4.2.10)
becomes

$$\langle (\varDelta n)^2 \rangle = \langle n \rangle,\qquad (4.2.15)$$

so that the photons behave as independent particles.

(iii) FOCK STATE   At first sight, we might be tempted to conclude from
(4.2.10) that $\langle \varDelta n)^2 \rangle \geqq \langle n \rangle$, i.e., that $\langle (\varDelta U)^2 \rangle \geqq 0$. However, we must
remember that the weight function φ in (4.2.5) is *not* positive semidefinite.
This can be seen clearly for a Fock state (which, as we have already observed, has no classical limit),

$$\varrho = |n\rangle \langle n|.\qquad (4.2.16)$$

This implies, of course,

$$\langle (\varDelta n)^2 \rangle = 0,\qquad (4.2.17)$$

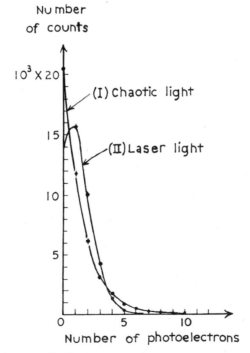

Figure 4.1   Photoelectron distribution for: (*I*) Chaotic ("thermal") light; (*II*) Laser
light (after F. T. Arecchi, *Phys. Rev. Letters* **15**, 912 (1965))

so that the corresponding $\varphi(v)$, which is a tempered distribution, containing finite-order derivatives of the delta function (as may readily be verified from (3.4.10), (3.4.13)), must be such that $\langle(\varDelta U)^2\rangle = -\langle n\rangle = -n$. It is, in fact, known that the derivatives of delta are not positive distributions.

The experimental investigation of the photon statistics is carried out through the corresponding photoelectron statistics, i.e., one tries to determine (for a stationary field) the probability $p(n, T)$ of counting n photoelectrons in a given time interval T. The experimental results obtained by Arecchi[5] for $p(n, T)$, with $T = 10^{-5}$ sec, are shown in Figure 4.1, for laser light and "thermal" light (actually, a chaotic source with properties similar to those of a thermal source but greater intensity). This time T is much smaller than the coherence time of the light employed, and, under those conditions, it can be shown[6] that $p(n, T)$ is essentially a measure of $p(n)$, the probability of finding n photons in the field. The curve for "thermal" light in Fig. 4.1 is in good agreement with a Bose distribution [cf. (4.1.8)], while that for laser light agrees well with a Poisson distribution [cf. (4.2.14)]. Thus, the two types of light have very different statistical properties, even though they might not be distinguished, e.g., by their spectral distribution.

4.3 HIGHER-ORDER COHERENCE EFFECTS

(a) Coherence functions for Gaussian light

It is a well known property of the Gaussian distribution (in any number of variables) that all of its higher-order moments can be evaluated in terms of the second-order moments[7]. It then follows from (4.1.15) that, for thermal light, all higher-order coherence functions can be expressed in terms of the second-order ones.

To verify this explicitly[8], let us extend the concept of normally-ordered characteristic function (3.4.44) by defining the *normally-ordered generating functional* $F_N[\zeta(x)]$,

$$F_N[\zeta(x)] = \mathrm{Tr}\left\{\varrho \exp\left[\int \zeta(x)E^{(-)}(x)d^4x\right] \exp\left[-\int \zeta^*(x)E^{(+)}(x)d^4x\right]\right\}. \quad (4.3.1)$$

We then have the result, analogous to (3.4.47),

$$G^{(n,n)}(x_1, \ldots, x_n; x_{n+1}, \ldots, x_{2n})$$

$$= \left[\frac{\delta^{2n}}{\delta\zeta(x_1) \ldots \delta\zeta(x_n) \delta[-\zeta^*(x_{n+1})] \ldots \delta[-\zeta^*(x_{2n})]} F_N\right]_{\zeta(x)=0}. \quad (4.3.2)$$

where $\delta/\delta\zeta$ denotes the functional derivative[9]; in fact, each functional differentiation brings down a factor $E^{(\pm)}$, and setting $\zeta = 0$ reduces the exponentials to unity, so that we get (2.2.31).

If we now employ the mode expansion (3.4.66),

$$E^{(+)}(x) = \sum_k f_k(x) \, a_k,$$ (4.3.3)

wo got

$$\int \zeta(x) \, E^{(-)}(x) \, d^4x = \sum_k \zeta_k a_k^+, \quad \zeta_k = \int \zeta(x) f_k^*(x) \, d^4x.$$ (4.3.4)

Substituting (3.4.64), (4.1.15) and (4.3.4) in (4.3.1), we get

$$F_N = \int \prod_k \frac{d^2 v_k}{\pi \langle n_k \rangle} \exp\left(-\frac{|v_k|^2}{\langle n_k \rangle}\right) \langle v_k| \, e^{\zeta_k a_k^+} \, e^{-\zeta_k^* a_k} |v_k\rangle$$

$$= \prod_k \frac{1}{\pi \langle n_k \rangle} \int \exp\left(-\frac{|v_k|^2}{\langle n_k \rangle}\right) \exp\left(\zeta_k v_k^* - \zeta_k^* v_k\right) d^2 v_k$$

$$= \prod_k \exp\left(-\langle n_k \rangle |\zeta_k|^2\right) = \exp\left(-\sum_k \langle n_k \rangle |\zeta_k|^2\right)$$

$$= \exp\left[-\sum_k \zeta_k \sum_{k'} \zeta_{k'}^* \, \mathrm{Tr}\left(\varrho a_k^+ a_{k'}\right)\right]$$

$$= \exp\left\{-\mathrm{Tr} \int d^4x \int d^4x' \, \zeta(x) \left[\varrho E^{(-)}(x) \, E^{(+)}(x')\right] \zeta^*(x')\right\},$$

where we have employed (4.1.12), (4.1.13), and the fact that

$$\mathrm{Tr}\left(\varrho a_k^+ a_{k'}\right) = \langle n_k \rangle \, \delta_{kk'},$$ (4.3.5)

because ϱ is stationary (φ depends only on $\{|v_k|\}$). Finally,

$$F_N[\zeta(x)] = \exp\left[-\int \int \zeta(x) \, G^{(1,1)}(x, x') \, \zeta^*(x') \, d^4x \, d^4x'\right],$$ (4.3.6)

so that all higher-order coherence functions are indeed determined by $G^{(1,1)}$.

Explicitly, in (4.3.2), each differentiation with respect to a pair of variables $[\zeta(x_i), -\zeta^*(x_j)]$ brings down a factor of the type $G^{(1,1)}(x_i, x_j)$. In view of the symmetry of (4.3.2) with respect to permutations of arguments in the same group, this leads to

$$G^{(n,n)}(x_1, \ldots, x_n; x_{n+1}, \ldots, x_{2n}) = \sum_P \prod_{i=1}^n G^{(1,1)}(x_i, x_{P(n+i)}),$$ (4.3.7)

where the sum is extended over all $n!$ permutations of the indices $n + 1$, ..., $2n$. For instance,

$$G^{(2,2)}(x_1, x_2; x_3, x_4) = G^{(1,1)}(x_1, x_3) \, G^{(1,1)}(x_2, x_4)$$
$$+ G^{(1,1)}(x_1, x_4) \, G^{(1,1)}(x_2, x_3).$$ (4.3.8)

Let us now discuss some applications of this result.

(b) Bunching effects

Consider a photodetector illuminated by a stationary light source. It is clear from (4.2.13) that, for thermal light, the counts will not be registered as if the photons were arriving at random (as independent classical particles): there must be correlations between the counts. A good measure of these correlations is the *conditional probability* $p_c(t \mid t + \tau)\,\Delta\tau$ of registering a count between $t + \tau$ and $t + \tau + \Delta\tau$, given that a count has already been registered at time t.

According to the discussion given in §2.2(c), this may be regarded as a kind of delayed coincidence experiment with a single counter at the same position $\mathbf{r}$, but two different times t, $t + \tau$, so that

$$p_c(t \mid t + \tau)\,\Delta\tau = \frac{s^2 G^{(2,2)}(t, t + \tau; t, t + \tau)\,\Delta t \Delta\tau}{s G^{(1,1)}(t, t)\,\Delta t}, \qquad (4.3.9)$$

where the denominator represents the probability for the first event to take place within Δt [cf. (2.2.25)], and we have omitted the dependence on position $\mathbf{r}$.

If we have *ideal laser light*, the field is fully coherent [§4.1(b)], so that, according to (2.4.17),

$$G^{(2,2)}(t, t + \tau; t, t + \tau) = G^{(1,1)}(t, t)\, G^{(1,1)}(t + \tau, t + \tau)$$

$$= [G^{(1,1)}(t, t)]^2 \qquad (4.3.10)$$

because of stationarity. Thus, in this case [cf. (2.2.25)]

$$p_c(t \mid t + \tau) = s G^{(1,1)}(t, t) = w, \qquad (4.3.11)$$

i.e., the conditional probability is the same as the unconditional one. There are no statistical correlations: the photons behave as independent particles. This agrees with the result (4.2.15) for the fluctuations in the number of photons.

On the other hand, if we have *thermal light*, it follows from (4.3.8) that

$$G^{(2,2)}(t, t + \tau; t, t + \tau) = G^{(1,1)}(t, t)\, G^{(1,1)}(t + \tau, t + \tau)$$

$$+ G^{(1,1)}(t, t + \tau)\, G^{(1,1)}(t + \tau, t) = [G^{(1,1)}(t, t)]^2\,[1 + |\gamma(\tau)|^2], \quad (4.3.12)$$

where we have employed (2.3.6) and (2.3.29). Thus, (4.3.9) yields

$$p_c(t \mid t + \tau) = w[1 + |\gamma(\tau)|^2], \qquad (4.3.13)$$

i.e., the correlation is a measure of the modulus of the *complex degree of temporal coherence* $\gamma(\tau)$. We always have [cf. (2.3.29)] $\gamma(0) = 1$; thereafter, $|\gamma(\tau)|$ tends to decrease with decay time given by the coherence time [cf. (1.3.36)]. Thus, immediately after the arrival of one photon, the probability of counting a second photon is twice as great as the probability after a long time: thermal photons show the well known "gregariousness" of bosons; they tend to stick together.

The experimental results obtained by Arecchi, Gatti and Sona[10] for laser light and light from an artificial "chaotic" source are shown in Fig. 4.2.

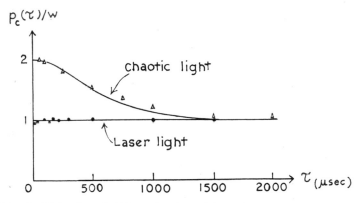

Figure 4.2 Conditional probability of a time delay τ between two successive photo-electric counts, for chaotic light and laser light [after F. T. Arecchi, E. Gatti, and A. Sona, *Phys. Letters* **20**, 27 (1966)]

(c) The Hanbury Brown and Twiss effect

We have seen in (1.3.43) that the Hanbury Brown and Twiss effect allows us to measure

$$\langle \Delta I(\mathbf{r}_1, t_1)\, \Delta I(\mathbf{r}_2, t_2) \rangle = \langle (I_1 - \langle I_1 \rangle)(I_2 - \langle I_2 \rangle) \rangle$$

$$= \langle I_1 I_2 \rangle - \langle I_1 \rangle \langle I_2 \rangle, \qquad (4.3.14)$$

where $I_1 = I(\mathbf{r}_1, t_1)$; $I_2 = I(\mathbf{r}_2, t_2)$. In terms of the quantum coherence functions, this is proportional to

$$\langle \Delta I_1\, \Delta I_2 \rangle \propto G^{(2,2)}(x_1, x_2; x_1, x_2) - G^{(1,1)}(x_1, x_1)\, G^{(1,1)}(x_2, x_2).$$

$$(4.3.15)$$

For laser light, according to (2.4.17), this would vanish.

On the other hand, for thermal light, (4.3.8) and (4.3.15) imply

$$\langle \Delta I_1\, \Delta I_2 \rangle \propto G^{(1,1)}(x_1, x_2)\, G^{(1,1)}(x_2, x_1) = |G^{(1,1)}(x_1, x_2)|^2, \qquad (4.3.16)$$

and, according to (3.4.68), this corresponds to $|\Gamma(\mathbf{r}_1, \mathbf{r}_2, \tau)|^2$, where $\tau = t_2 - t_1$ and Γ is defined by (1.3.14). This proves the result mentioned in § 1.3(d): for thermal light, the Hanbury Brown and Twiss effect allows us to measure the modulus of the mutual coherence function between two points of the field.

(d) Interference of independent beams

Traditionally, optical interference experiments have been performed by splitting a light beam from a single source into two or more components, and then reuniting them with some optical path differences, as is done, e.g., in the Michelson interferometer (§ 1.2). What happens if we superpose light beams from two independent sources?

Classically, such light beams may always interfere: if they are quasi-monochromatic, the corresponding wave functions are nearly sinusoidal for times much shorter than the coherence times of the two beams [§ 1.3(c)]. The superposition gives rise to transient interference fringes, with the whole interference pattern shifting and changing in time as a function of the phase difference between the beams.

If one wants to describe this effect in quantum-mechanical terms, especially for very weak light, there seem to arise some difficulties of interpretation, associated with Dirac's[11] well known statement: "Each photon ... interferes only with itself. Interference between two different photons never occurs." If the light sources are so weak that one is detecting essentially just one photon at a time, does this imply that beams from independent sources will not interfere?

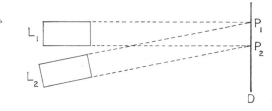

Figure 4.3 Experimental arrangement for observing transient interference fringes. L_1, L_2: lasers; D: detector

In order to perform the experiment, one must employ laser sources. In fact, the fringes can only appear over an area comparable with the coherence area (§ 1.2), and they have to be observed in a time interval smaller than the coherence time, so that the average number of photons observed is of the order of the degeneracy parameter δ; as we have seen in § 1.2, $\delta \ll 1$ for thermal light, but it can become very large for laser light.

The first observation of transient interference fringes with beams from two independent lasers was made by Magyar and Mandel[12]. The experimental arrangement is schematically indicated in Fig. 4.3: the beams from the two lasers are directed at a small angle to each other upon a detector (e.g., a photographic plate). The transient interference fringes were observed, in this case, with intense light beams. The experiment was repeated with highly attenuated beams by Pfleegor and Mandel[13]. The transit time between sources and detectors was about 3 nsec, and the light was so weak that the mean interval between arrivals of photons was about 150 nsec, so that one could say with high probability that each photon was absorbed before the next one was emitted. A special type of detector had to be employed. The results demonstrated that the transient interference fringes are still present under these conditions.

In order to give a quantum treatment of the transient interference effects, let us assume, to begin with, that during time intervals comparable with those of the observations, the field produced by each laser may be described by a pure coherent state, as in (4.1.19); say, $|\{v_k\}\rangle$ for one of them and $|\{v'_k\}\rangle$ for the other one. We assume that the fields due to the two sources have no modes in common (because the two beams are inclined at an angle to each other). Since the modes are independently excited, it then follows that the resultant state vector for the combined field may be described by the direct product $|\{v_k\}\rangle\,|\{v'_k\}\rangle$.

Let
$$E^{(+)}(x)\,|\{v_k\}\rangle = V(x)\,|\{v_k\}\rangle, \tag{4.3.17}$$
$$E^{(+)}(x)\,|\{v'_k\}\rangle = V'(x)\,|\{v'_k\}\rangle, \tag{4.3.18}$$
as in (3.1.2). Then,
$$E^{(+)}(x)\,|\{v_k\}\rangle\,|\{v'_k\}\rangle = [V(x) + V'(x)]\,|\{v_k\}\rangle\,|\{v'_k\}\rangle, \tag{4.3.19}$$
i.e., the coherent states are added as classical amplitudes.

The probability of a photon absorption at the point x, according to (2.2.25), is then proportional to
$$\begin{aligned} G^{(1,1)}(x, x) &= \langle\{v'_k\}|\,\langle\{v_k\}|\,E^{(-)}(x)\,E^{(+)}(x)\,|\{v_k\}\rangle\,|\{v'_k\}\rangle \\ &= |V(x)|^2 + |V'(x)|^2 + 2\,\mathrm{Re}\,[V^*(x)\,V'(x)], \end{aligned} \tag{4.3.20}$$
and it does contain an interference term.

However, we are making the very strong assumption that the phase of the laser field is completely known, so that it is in a pure state. In practice, as remarked in § 4.2(b), we usually have no phase information, so that it is more realistic to employ a density operator corresponding to a mixture, such as (4.1.23), which describes a stationary field. This corresponds to averaging over the phases, so that the interference term in (4.3.20) is wiped out in the statistical (ensemble) average.

This does *not* mean, however, that the interference fringes do not appear in any individual experiment. We may associate each member of the ensemble, as in (4.1.19), with a definite phase, so that the fringes appear in

6*

each single trial. Their position, however, varies from one trial to the next, and when we perform the ensemble average, we are taking an average over all transient interference patterns, in which the fringes are wiped out.

In order to observe the fringes, one measures the intensities at several different points. However, as soon as one observes the intensities at a couple of points in one experiment, the phase difference between the two lasers is determined. We can then describe the remainder of the pattern in terms of a reduced density operator which incorporates this new knowledge.

Alternatively, as was actually done in the Pfleegor-Mandel experiment, we can detect the interference fringes in terms of the correlations between intensities at different points, such as P_1 and P_2 in Fig. 4.3. They give rise to oscillatory terms in the fourth-order coherence function $G^{(2,2)}(x_1, x_2; x_1, x_2)$; the treatment is rather similar to that of the Hanbury Brown and Twiss effect.

Finally, let us note that the above discussion is still consistent with Dirac's statement. In fact, as we have seen in (4.3.20), the interference effects already appear in the absorption of *one* photon. Similarly, in the intensity correlations (as well as in the Hanbury Brown and Twiss effect), the interferences are due to the fact that one cannot tell from which of the two sources the photon is coming[13a]. Thus, in Fig. 4.3, there are two possibilities: $(L_1 \to P_1)$, $(L_2 \to P_2)$, and the "crossed term" $(L_1 \to P_2)$, $(L_2 \to P_1)$. It is the probability amplitudes (Feynman "histories") corresponding to these two possible paths that interfere. If the experimental arrangement were modified in such a way that one could tell from which source each photon was originating, the interference pattern would disappear. In this sense, it is indeed correct to say that "different photons do not interfere".

4.4 THE RECONSTRUCTION THEOREM[14]

Let us now go back to the problem that was raised at the end of § 2.3: given the set of all coherence functions $G^{(n,m)}$ (for all values of n and m), can one reconstruct the density operator of the field?

It is sufficient to restrict the discussion to the case of a single mode; the general case can be reduced to this one by taking a direct product over the modes [§ 3.4(d)]. As we have seen in § 3.4(c), the density operator is completely determined if we are given the normally-ordered characteristic function $C_N(z)$ for all values of z.

In fact, according to (3.4.28), (3.4.53), and (3.4.54), we have

$$\langle q, p| \varrho |q, p \rangle = \frac{1}{2\pi} \int e^{i(xp-kq)} \exp\left[-\tfrac{1}{2}(x^2 + k^2)\right] C_N\left(\frac{x + ik}{\sqrt{2}}\right) dx\, dk.$$

$$(4.4.1)$$

Since ϱ is uniquely characterized by its diagonal matrix elements, we see that it is uniquely determined by $C_N(z)$. Actually, according to (3.4.51), the weight function $\varphi(v)$ of the diagonal representation is nothing but the Fourier transform of $C_N(z)$, so that the density operator can, in principle, be explicitly reconstructed, given $C_N(z)$ (this is to be understood in the symbolic sense of § 3.4(b) in singular cases).

On the other hand, given $G^{(n,m)}$ [cf. (3.4.68)], we may determine their Fourier transforms $\tilde{G}^{(n,m)}$ defined by (3.4.70) [equivalently, we may expand the field in terms of the set of modes (2.4.22)]. For a single mode, $\tilde{G}^{(n,m)}$ is given by (3.4.47).

The normally-ordered characteristic function is related with the given set of functions $\tilde{G}^{(n,m)}$ by (3.4.46):

$$C_N(z) = \sum_{n=0}^{\infty} \sum_{m=0}^{\infty} \frac{\tilde{G}^{(n,m)}}{n!m!} z^n(-z^*)^m. \tag{4.4.2}$$

Therefore, if this series converges for all z, we can employ it to compute $C_N(z)$ and thus solve the reconstruction problem.

We conclude that *a sufficient condition for the reconstruction to be possible is that*

$$C_N(z, w) = \sum_{n=0}^{\infty} \sum_{m=0}^{\infty} \frac{\tilde{G}^{(n,m)}}{n!m!} z^n w^m \tag{4.4.3}$$

be an entire analytic function of the two independent complex variables z and w. Note that this is not a *necessary* condition: even if the series has a finite domain of convergence, it may still be possible to obtain $C_N(z, w)$ outside of it by analytic continuation, and all we need is the value of this function for $w = -z^*$. According to the theorem quoted following (3.3.22), $C_N(z, w)$ is already determined by the values of (4.4.2) in a neighborhood of $z = 0$.

According to a theorem on power series expansions of functions of several complex variables[15], *a necessary and sufficient condition for (4.4.3) to define an entire function in both variables is that*

$$\lim_{m+n \to \infty} \left\{ \left[\frac{|\tilde{G}^{(n,m)}|}{n!m!} \right]^{\frac{1}{m+n}} \right\} = 0. \tag{4.4.4}$$

Thus, if the asymptotic behavior of $|\tilde{G}^{(n,m)}|$ for $n + m \to \infty$ is such that (4.4.4) is satisfied, the reconstruction of ϱ is possible by the above procedure.

It follows from (3.4.10), (3.4.53), and (3.4.54) that

$$C_N(z) = e^{z \cdot z^*} \cdot \frac{1}{\pi} \int e^{zv^* - z^*v} \langle v| \varrho |v\rangle d^2v. \tag{4.4.5}$$

This allows us to employ the representation

$$C_N(z, w) = e^{-zw} \cdot \frac{1}{\pi} \int e^{zv^* + wv} \langle v| \varrho |v\rangle \, d^2v \qquad (4.4.6)$$

in order to investigate the analytic properties in z and w.

In particular, let us assume that

$$\langle v| \varrho |v\rangle \leq e^{-\varepsilon |v|^2}, \qquad \varepsilon > 0, \qquad (4.4.7)$$

and let us consider the domain

$$|z| \leq R, \qquad |w| \leq R. \qquad (4.4.8)$$

Then, it follows from (4.4.6) that, for large R,

$$|C_N(z, w)| \leq \frac{1}{\pi} e^{R^2} \int e^{2R|v| - \varepsilon|v|^2} |v| \, d\,|v| \, d(\arg v)$$

$$\lesssim \frac{2R}{\pi^{1/2} \varepsilon^{3/2}} \exp\left[\left(1 + \frac{1}{\varepsilon}\right) R^2\right]. \qquad (4.4.9)$$

The quantity $\langle v| \varrho |v\rangle$ is the statistical expectation value of the projector $|v\rangle \langle v|$ on the coherent state $|v\rangle$; it is also normalized according to (3.4.15). We can interpret it [cf., however, § 3.4(e)] as representing a kind of "probability of finding the system in the coherent state $|v\rangle$", and we have, in any case,

$$0 \leq \langle v| \varrho |v\rangle \leq 1. \qquad (4.4.10)$$

All that matters, then, for the validity of the above estimates, is the asymptotic behavior as $|v| \to \infty$, i.e., we may replace (4.4.7) by

$$\langle v| \varrho |v\rangle = O(e^{-\varepsilon|v|^2}), \qquad |v| \to \infty \quad (\varepsilon > 0). \qquad (4.4.11)$$

The integrand of (4.4.6) is an entire function of z and w; under the above conditions, the integral is absolutely and uniformly convergent. Therefore, it represents an entire analytic function of z and w. The bound (4.4.9) as $R \to \infty$ implies that it is an entire function of order ≤ 2 [this is consistent with (3.4.58)] and type[16] $\leq 1 + 1/\varepsilon$. We, therefore, arrive at the following theorem:

If the "probability of finding the field in a coherent state $|v\rangle$" decreases at least exponentially with the average number of photons $\langle n\rangle = |v|^2$ contained in $|v\rangle$ as $\langle n\rangle \to \infty$, so that (4.4.11) is satisfied, it follows that $C_N(z, w)$ is an entire function of z and w, of order ≤ 2 and type $\leq 1 + 1/\varepsilon$.

Let us illustrate these results with some examples. For a *Fock state* with N photons, or for a superposition of such states with occupation numbers

$\leqq N$, all $\tilde{G}^{(n,m)} = 0$ for $n \geqq N$ or $m \geqq N$ [cf. (2.3.4)], so that (4.4.4) is obviously satisfied, and $C_N(z, w)$ is a polynomial. By (3.1.26), we have, for $\varrho = |N\rangle \langle N|$,

$$\langle v| \varrho |v\rangle = \frac{|v|^{2N}}{N!} \exp\left(-|v|^2\right),$$

so that (4.4.11) is satisfied, with $\varepsilon = 1$.

For a *coherent state* $\varrho = |v_0\rangle \langle v_0|$, we have

$$\tilde{G}^{(n,m)} = (v_0^*)^n v_0^m,$$

which satisfies (4.4.4). Substituting in (4.4.3), we get

$$C_N(z, w) = \exp\left(zv_0^* + wv_0\right),$$

whereas, by (3.2.28),

$$\langle v| \varrho |v\rangle = |\langle v | v_0\rangle|^2 = \exp\left(-|v - v_0|^2\right),$$

which satisfies (4.4.11), with $\varepsilon = 1$. In this case, $C_N(z, w)$ is an entire function of exponential type, so that the theorem is satisfied. A superposition of a finite number of coherent states may be treated in a similar manner.

For a *thermal field*, it follows from (4.1.14) and (3.4.47) that

$$\tilde{G}^{(n,m)} = n! \langle n\rangle^n \delta_{n,m},$$

and we find, with the help of Stirling's formula, that (4.4.4) is still satisfied. Substituting in (4.4.3), we get

$$C_N(z, w) = \exp\left(\langle n\rangle zw\right),$$

and (3.4.28), (4.1.11) and (3.4.12) yield

$$\langle v| \varrho |v\rangle = \frac{1}{1 + \langle n\rangle} \exp\left(-\frac{|v|^2}{1 + \langle n\rangle}\right).$$

It follows that (4.4.11) is satisfied, with $\varepsilon = (1 + \langle n\rangle)^{-1}$, so that, according to the above theorem, $C_N(z, w)$ is of order $\leqq 2$ and type $\leqq 2 + \langle n\rangle$. The explicit expression above shows that it actually is of order 2 and type $\langle n\rangle$, so that the theorem gives a very good result.

Thus, in all these cases, which represent the best-known models in quantum optics, the density operator can be explicitly reconstructed if we are given the set of all coherence functions $G^{(n,m)}$.

This shows that the coherence functions, although their definition appears to be linked with the somewhat incidental fact that ordinary detectors operate by absorption [§ 2.2(c)], actually have a much more fundamental significance: they contain all the statistical information about the field.

References

1. Cf. R. J. Glauber, *Phys. Rev.* **131**, 2766 (1963).
2. R. J. Glauber, *Phys. Rev.* **84**, 395 (1951); *QOE* (see Ch. 1, Ref. 2), p. 132. Cf., also, T. W. B. Kibble, *J. Math. Phys.* **9**, 315 (1968).
3. L. Mandel, *Proc. Phys. Soc. (London)* **72**, 1037 (1958).
4. A. Einstein, *Phys. Z.* **10**, 185 (1909).
5. F. T. Arecchi, *Phys. Rev. Letters* **15**, 912 (1965).
6. Cf. L. Mandel and E. Wolf, *Rev. Mod. Phys.* **37**, 231; 269 (1965).
7. Cf., e.g., C. L. Mehta, in *Lectures in Theoretical Physics*, ed. by W. E. Brittin, Univ. of Colorado Press, Boulder (1965), p. 345.
8. Cf. R. J. Glauber, *QOE*, p. 148.
9. Cf., e.g., L. I. Schiff, *Quantum Mechanics*, 2nd ed., McGraw-Hill, New York (1955), p. 332.
10. F. T. Arecchi, E. Gatti and A. Sona, *Phys. Letters* **20**, 27 (1966).
11. P. A. M. Dirac, *Quantum Mechanics*, 3rd ed., Oxford Univ. Press (1947), p. 9.
12. G. Magyar and L. Mandel, *Nature* **198**, 255 (1963).
13. R. L. Pfleegor and L. Mandel, *Phys. Rev.* **159**, 1084 (1967); cf., also, L. Mandel, *QO* (see Ch. 1, Ref. 3), p. 176.
13a. Cf., also, L. Menegozzi, *Suppl. Nuovo Cimento* **4**, 15 (1965).
14. The results presented in this section, as well as some of those given in § 2.3, are based on unpublished work by the author.
15. B. A. Fuks, *Introduction to the Theory of Analytic Functions of Several Complex Variables*, American Mathematical Society, Providence (1963), p. 51.
16. For the definitions of order and type, cf. B. Fuks, ibid., p. 339.

CHAPTER 5

The Laser

5.1 INTRODUCTION

THE EXTENSION OF the maser principle to the optical domain was first
proposed by Schawlow and Townes[1]. The first laser (the word is an acronym
for *light amplification by stimulated emission of radiation*) was built by
Maiman[2]. It was a ruby laser, with the light output in the form of pulses.
The first CW („continuous wave", i.e., producing a continuous light beam)
laser was the He-Ne gas laser, due to Javan[3]. A great variety of other laser
systems, utilizing solids (including semi-conductors), liquids and gases and
generating light over a frequency domain ranging from the infrared to the
ultraviolet, have been developed since.

Our aim here is to discuss in the simplest possible terms the theory of
the laser, so that we will not attempt to give any description of the various
types of lasers and of their manifold applications[4]. We will confine our
attention exclusively to *gas lasers*.

Basically, in the laser, an electromagnetic field at resonance with an
atomic transition between two levels gives rise to stimulated emission: the
emitted radiation is still at resonance, so that it gives rise to further transitions in other atoms, and in this way a kind of chain reaction or photon
avalanche can develop. A prerequisite, of course, is that the upper level
must be sufficiently populated; as this is not normally true at thermal

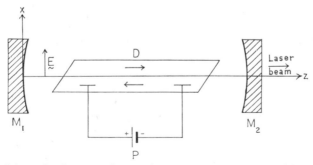

Figure 5.1 Schematic diagram of a gas laser. D: discharge tube with Brewster-angle
windows, which select the polarization with the electric field $\mathbf{E}$ as shown; P: power supply;
M_1, M_2: spherical mirrors

89

equilibrium, the population inversion must be achieved by artificial means (this is called *pumping*).

Another important requirement has to do with the resonance conditions. In contrast with a maser, a resonant cavity in the optical domain would have an enormously large number of modes competing for the energy in the frequency range of the atomic transition. In order to attain laser action, this has to be reduced to a relatively small number. To achieve this one employs, instead of a cavity, a laterally open system with mirrors at the ends, as will be explained below.

A typical He-Ne laser is shown schematically in Fig. 5.1. A *dc* (or *rf*) discharge is maintained in a discharge tube which contains, typically, a mixture of He at a pressure of 1 mm/Hg and Ne at 0.1 mm/Hg. The Ne atoms constitute the active medium; the He atoms are responsible for the pumping.

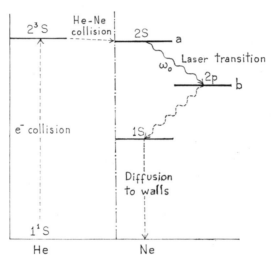

Figure 5.2 Schematic representation of part of the He-Ne level diagram, showing the $2S \rightarrow 2p$ 1.1μ infrared laser transition. Actually, the excited states of Ne are groups of closely spaced levels, instead of single levels as shown

The primary source of energy are the electrons accelerated in the discharge, which collide with He atoms and excite them to various states. The level diagram for He is shown in Fig. 5.2; the ground state is the singlet state 1^1S (parhelium). By electronic collisions, many He atoms end up in the excited state 2^3S (orthohelium), from which a direct transition to the ground state is forbidden, so that this is a long-lived metastable state. The basic Ne levels a and b between which the laser action takes place are the $2S$ and $2p$ levels. The $2S$ level lies only slightly below the 2^3S level of He, so that the pumping action takes place by collisions of the second kind between excited He atoms and Ne atoms in the ground state, in which there is a resonant transfer of the excitation from He to Ne (the small excess is taken up by kinetic energy).

The laser transition $2S \rightarrow 2p$, which gives rise to 1.1μ infrared radiation, has a spontaneous emission lifetime of 10^{-7} sec. It is followed by a much more rapid (lifetime 10^{-8} sec) transition to the $1S$ level, which decays to the ground state by collisions with the walls (or with impurities). A similar mechanism is responsible for the widely employed red line at 6328 Å. The fact that both a and b are excited states makes it much easier to achieve population inversion. Actually, both a and b in Fig. 5.2 contain several sublevels, but we will ignore this and consider single (nondegenerate) levels.

5.2 OPTICAL RESONATOR MODES

A resonant cavity of the type shown in Fig. 5.3 would have a set of resonant frequencies

$$\Omega^2_{nlm}/c^2 = k^2_{z,n} + k^2_{T,lm}, \tag{5.2.1}$$

where

$$k_{z,n} = n\pi/L, \tag{5.2.2}$$

$$k^2_{T,lm} = k^2_{x,l} + k^2_{y,m}; \quad k_{x,l} = l\pi/a, \quad k_{y,m} = m\pi/a, \tag{5.2.3}$$

where (l, m, n) are integers.

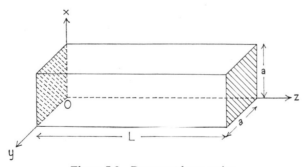

Figure 5.3 Resonant laser cavity

If we now open up the lateral walls (or make them transparent), leaving mirrors only at the ends, we no longer have, strictly speaking, a discrete set of modes: radiation can escape by the sides. Each mode acquires a "lifetime" or "width", and when the widths become too broad, they merge into a continuous spectrum. A proper treatment of such a problem would involve computing the poles of the scattering matrix associated with the system; their imaginary part would then give the width of the corresponding mode[5].

The longest-lived modes are the "axial" ones, corresponding to $l = m = 0$, associated with the normalized mode functions

$$u_n(z) = \sqrt{\frac{2}{L}} \sin (k_{z,n}z) = \sqrt{\frac{2}{L}} \sin (n\pi z/L) \quad (n = 1, 2, ...), \quad (5.2.4)$$

which correspond to waves traveling perpendicularly between the end mirrors (the exact modes will also depend on x and y, but we expect that this dependence will not be strong near the axis). Typically, $L \sim 10^2$ cm, so that optical frequencies correspond to $n \sim 10^6$. The frequency separation between two consecutive axial modes $(n, n + 1)$ is of the order 10^2 Mc.

The "lateral" modes correspond to nonvanishing l or m; typically, $a \sim 1$ cm, and the frequency separation between an axial mode and the nearest lateral one is ~ 1 Mc. Thus, we see that there is indeed a considerable clustering of lateral modes, and the main purpose of opening up the lateral walls is to get rid of them. In fact, they have wave vectors with lateral components, so that they can escape by the sides, and their lifetime should become very short as l and m increase.

The axial modes also have a finite lifetime. In fact, in order to extract the laser beam, we have to make one of the mirrors (e.g., M_2 in Fig. 5.1) partially transparent. If the corresponding transmissivity is α (usually $\ll 1$), the intensity is attenuated by a factor

$$(1 - \alpha)^N \approx e^{-N\alpha}$$

after N double traversals of the laser, corresponding to a *mode lifetime*

$$\tau_n = \frac{2L}{c\alpha}. \tag{5.2.5}$$

The transmission through the end mirror is not the only factor contributing to α. There are also *diffraction losses*, due to the fact that, after each reflection, a fraction of the beam is not intercepted by the opposite mirror, and, therefore, escapes through the sides. Optical resonators are usually designed to minimize the diffraction losses for the axial modes of interest (at the same time, of course, the losses are large for lateral modes). The usual treatment of such systems[6] is based on classical diffraction theory. A configuration often employed (Fig. 5.1) corresponds to two confocal spherical mirrors (each spherical surface is centered at the other one); it yields very low diffraction losses for axial modes.

It is customary to express the mode lifetime (5.2.5) in terms of a corresponding "Q" factor, defined by

$$Q_n = \Omega_n/\text{fraction of stored energy dissipated per unit time.}$$

Since the energy would decay like $\exp(-t/\tau_n)$, we have

$$Q_n = \Omega_n \tau_n = \Omega_n/\Delta\Omega_n = \nu_n/\Delta\nu_n, \qquad (5.2.6)$$

where $\Delta\nu_n$ is the mode frequency width. From (5.2.2), (5.2.5) and (5.2.6), we find

$$Q_n = 2n\pi/\alpha. \qquad (5.2.7)$$

For the values of n already mentioned and $\alpha \sim 2\%$, we get $Q_n \sim 10^8$. The width of resonant cavity modes is of the order of 1 Mc.

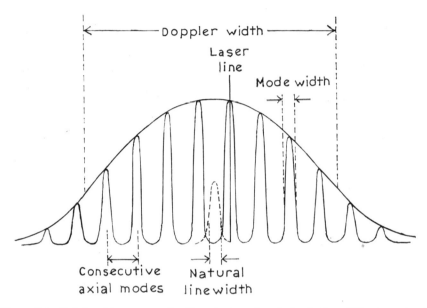

Figure 5.4 Relationships among the Doppler-broadened atomic linewidth, the natural linewidth and the width of axial modes in a typical gas laser. The laser output line is practically monochromatic

The natural linewidth for the $2S \rightarrow 2p$ transition in Fig. 5.2 is also of the order of 1 Mc, but the actual width in the gas discharge is of the order of 900 Mc, due to Doppler broadening. Since the spacing between consecutive axial modes is of order 10^2 Mc, one can still have several modes falling within the atomic linewidth. The typical situation is represented in Fig. 5.4. Unless special precautions are taken, several modes will be excited and compete among themselves for the energy released from the atomic excitation. The modes with highest Q and located near the center of the atomic line may be expected to grow more rapidly.

5.3 ELEMENTARY THEORY OF THE THRESHOLD

The critical population inversion density that defines the threshold above which laser action takes place can be evaluated very simply by computing the rate at which field energy is gained from and lost to the atoms of the active medium. The threshold is attained when the net gain from the atoms balances the mode loss described in § 5.2.

Let us consider the propagation of a light beam in a medium with N_a active atoms per unit volume in the upper state a and N_b in the lower state b. Let W_{ab} be the transition probability per unit time from a to b due to *induced emission*, and W_{ba} the transition probability per unit time from b to a due to *absorption*. Since each emission at frequency v increases the field energy by hv and each absorption decreases the energy by the same amount, the change dI_v in the beam intensity (energy per unit time and area) at frequency v after going through a distance dz in the medium is given by

$$dI_v = (N_a W_{ab} - N_b W_{ba}) \, hv \cdot dz. \tag{5.3.1}$$

As is well known[7], the above transition probabilities are proportional to the energy density $u(v)$ of the radiation already present at frequency v, and the proportionality constants are Einstein's B coefficients:

$$W_{ab} = B_{ab} u(v), \qquad W_{ba} = B_{ba} u(v). \tag{5.3.2}$$

From Einstein's well-known arguments about atoms in thermal equilibrium with blackbody radiation, it follows that

$$B_{ab} = B_{ba} \quad \text{(detailed balance)}, \tag{5.3.3}$$

$$B_{ab} = \frac{c^3}{8\pi h v^3} A_{ab}, \tag{5.3.4}$$

where Einstein's A coefficient A_{ab} is the probability per unit time for *spontaneous emission*, i.e.,

$$A_{ab} = \frac{1}{\tau_{ab}}, \tag{5.3.5}$$

where τ_{ab} is the lifetime for spontaneous emission.

These results are valid not just for blackbody radiation, but also for any field in which $u(v)$ is essentially constant over the frequency range that is effective in inducing atomic transitions. However, as we see from Fig. 5.4, this is by no means true for the laser: the width of each field mode is much smaller than the effective atomic linewidth, and the effectiveness of the mode in inducing transitions will depend on its relative location within the atomic linewidth.

To account for this, we introduce the *normalized atomic line shape factor* $g(v)$, with $g(v) \, dv$ representing the probability that a given transition will correspond to the emission (or absorption, since the shape of the absorption line is the same as that of the emission line) of a photon with frequency between v and $v + dv$, so that

$$\int g(v) \, dv = 1. \tag{5.3.6}$$

The relation (5.3.2) is then replaced by

$$W_{ab} = W_{ba} = B_{ab} u(v) \, g(v) = \frac{c^3 u(v) \, g(v)}{8\pi h v^3 \tau_{ab}}, \tag{5.3.7}$$

where (5.3.3) and (5.3.4) have also been employed. In particular, if $u(v)/v^3$ does not change appreciably across the atomic linewidth, the total transition rate for induced emission is obtained by integrating over v, and we recover (5.3.2), with the help of (5.3.6).

The natural line shape for the atomic transition is described by the normalized Lorentzian line shape factor,

$$g(v) = \frac{\Delta v}{2\pi} \, \frac{1}{[(v - v_0)^2 + (\Delta v/2)^2]}, \tag{5.3.8}$$

where v_0 is the transition frequency and $\Delta v = \Delta \omega / 2\pi$, the width at half-maximum, corresponds to $(2\pi\tau_{ab})^{-1}$. The damping due to spontaneous emission is a line broadening effect which acts in an identical way for every atom, so that it is called *homogeneous broadening*.

As indicated in Fig. 5.4, the actual width of the atomic line in a gas is much greater than the natural linewidth. There are two main factors responsible for this:

(a) *Atomic collisions* interrupt the emission process from an excited atom, shortening the lifetime and, thereby, giving rise to *collision broadening*. This effect can still be described by the Lorentzian lineshape function (5.3.8), by enlarging the width Δv to take account of the shortened lifetime.

(b) *The Doppler effect* An atom moving with velocity v in the z-direction gives rise to a Doppler frequency shift $(v - v_0)/v_0 = v/c$. Taking into account the Maxwellian velocity distribution of the atoms, this gives rise to a frequency distribution proportional to $\exp\left[-\beta mc^2 \, \frac{(v - v_0)^2}{2v_0^2}\right]$,

where $\beta = 1/\varkappa T$, as in (2.3.14). Since this broadening effect arises from the fact that different atoms have different velocities, and therefore contribute at different frequencies [rather than from a homogeneous damping effect,

as (5.3.8)], it is called *inhomogeneous broadening*. The corresponding normalized line shape function is given by

$$g(\nu) = \frac{2(\ln 2)^{\frac{1}{2}}}{\pi^{\frac{1}{2}} \Delta\nu} \exp\left[-4\ln 2 \frac{(\nu - \nu_0)^2}{(\Delta\nu)^2}\right], \qquad (5.3.9)$$

where

$$\Delta\nu = \Delta\nu_D = 2\nu_0 \left(\frac{2kT\ln 2}{mc^2}\right)^{\frac{1}{2}} \qquad (5.3.10)$$

is the Doppler width. In general, both collision broadening and Doppler
broadening have to be taken into account, and the resulting line shape
function is somewhat more complicated.

The intensity I_ν is related with the energy density $u(\nu)$ by

$$I_\nu = cu(\nu), \qquad (5.3.11)$$

where we are approximating the refractive index of the gas by unity. Substituting (5.3.11) and (5.3.7) in (5.3.1), we find

$$\frac{1}{I_\nu} \frac{dI_\nu}{dz} = \frac{c^2 g(\nu) N}{8\pi\nu^2 \tau_{ab}} = \beta(\nu), \qquad (5.3.12)$$

where

$$N = N_a - N_b \qquad (5.3.13)$$

is the *density of population inversion*.

Integrating (5.3.12) with respect to z, we get

$$I_\nu(z) = I_\nu(0) \, e^{\beta(\nu)z}. \qquad (5.3.14)$$

If we have no pumping, i.e., for an ordinary gas at thermal equilibrium, the
population densities N_a and N_b are related by a Boltzmann factor, so that
$N_a \ll N_b$, and $N \approx -N_b$. In this case, (5.3.14) describes the exponential
attenuation of the beam as it propagates through the medium, due to
absorption, and $\beta(\nu)/N_b$ represents simply the *absorption cross-section* per
atom.

The first condition for laser amplification to be possible is that a *population inversion* be produced by pumping, i.e., that $N > 0$. Under these conditions, (5.3.14) represents an exponential *growth* with distance. This is
not sufficient, however, because we must compensate for the energy losses
discussed in § 5.2.

For each double traversal of the laser, (5.3.14) leads to a gain factor of
$e^{2\beta(\nu)L}$, whereas, according to § 5.2, the effect of the losses is represented
by a damping factor $e^{-\alpha}$. The overall gain factor is therefore

$$G(\nu) = e^{2\beta(\nu)L - \alpha}, \qquad (5.3.15)$$

and the condition for laser oscillation to be possible at frequency ν is that

$$G(\nu) \geq 1. \qquad (5.3.16)$$

According to (5.3.12), (5.2.5), and (5.2.6), this is equivalent to

$$\beta(v) = \frac{c^2 g(v)\, N}{8\pi v^2 \tau_{ab}} \geqq \frac{\alpha}{2L} = \frac{1}{c\tau_n} = \frac{\Omega_n}{cQ_n}. \qquad (5.3.17)$$

Other things being equal, the maximum gain factor is obtained by taking the maximum for $g(v)$, which corresponds to the center frequency of the atomic line, $v = v_0$. The corresponding density of population inversion, given by (5.3.17) for $v = v_0$, is called the *critical density of population inversion* N_c, and it defines the *threshold condition* for laser oscillation at the resonance frequency v_0:

$$N_c = \frac{8\pi v_0^2}{c^3 g(v_0)} \frac{\tau_{ab}}{\tau_n}, \qquad (5.3.18)$$

where τ_n is the lifetime of the mode with angular frequency $\Omega_n = \omega_0 = 2\pi v_0$, i.e., at resonance with the atomic transition frequency. The maximum value $g(v_0)$ of the atomic lineshape factor is inversely proportional to the atomic linewidth Δv. In fact, according to (5.3.8) and (5.3.9),

$$g(v_0) = \frac{2}{\pi \Delta v} \qquad \text{(Lorentzian)}, \qquad (5.3.19)$$

$$g(v_0) = \frac{2(\ln 2)^{\frac{1}{2}}}{\pi^{\frac{1}{2}} \Delta v} \qquad \text{(Doppler)}. \qquad (5.3.20)$$

We see from (5.3.18) that, in order to render the critical population inversion as low as possible, the following requirements should be satisfied:

(a) The spontaneous emission lifetime τ_{ab} should be as small as possible. According to (5.3.7), this increases the transition rate for induced emission.
(b) For the same reason, $g(v_0)$ should be large, i.e., the atomic line width Δv should be as small as possible. As seen in Fig. 5.4, this also reduces the number of field modes falling within the atomic line width. (c) The mode lifetime τ_n (or, equivalently, its "Q" factor Q_n), should be as large as possible, in order to reduce the losses.

Some typical numerical values for a He-Ne laser are: $\tau_{ab} \sim 10^{-7}$ sec; $v_0 \sim 3 \times 10^{14}$ cps (for the 1.1μ line); Δv (Doppler width) $\sim 9 \times 10^8$ cps; $\tau_n \sim 3 \times 10^{-7}$ sec (for $L = 1$ m, $\alpha \sim 10^{-2}$). Substituting these values in (5.3.18), we find for the critical inversion density: $N_c \sim 2 \times 10^7$ atoms/cm^3. Since the total density of Ne atoms in the discharge, at a pressure of 0.1 mm/Hg, is of the order of 10^{16} atoms/cm^3, note that *the active atoms constitute a highly rarefied gas*. In order to produce the population inversion, the minimum power required per unit volume is of the order of $N_c hv_0/\tau_{ab}$; for a laser volume of 100 cm^3, this is about 7 mw.

7   Nussenzveig (0380)

Beyond the threshold, according to (5.3.15), the intensity would grow exponentially. Actually, of course, the total output energy cannot exceed the energy supplied by pumping, but this *saturation* effect, which would imply a decrease in the inversion density N, does not appear in the above treatment, where N is assumed to be given. Saturation is a nonlinear effect; this already shows that nonlinearities play an essential role in stabilizing the intensity. Although a linear treatment suffices to determine the threshold condition, the steady-state solution, corresponding to the stable amplitude of oscillation above threshold, can only be obtained from a nonlinear theory.

References

1. A. L. Schawlow and C. H. Townes, *Phys. Rev.* **112**, 1940 (1958).
2. T. H. Maiman, *Nature* **187**, 493 (1960).
3. A. Javan, W. B. Bennett, Jr., and D. R. Herriott, *Phys. Rev. Letters* **6**, 106 (1961).
4. Cf. A. Yariv and J. P. Gordon, *Proc. IEEE* **51**, 4 (1963); B. A. Lengyel, *Lasers*, 2d ed., Wiley, New York (1971); Smith and Sorokin, *The Laser*, McGraw-Hill, New York (1966); A. Yariv, *Quantum Electronics*, Wiley, New York (1968); F. T. Arecchi and E. O. Schulz-Dubois, eds., *Laser Handbook*, Vols. 1 and 2, North-Holland Publishing Co., Amsterdam (1972).
5. Cf. G. Beck and H. M. Nussenzveig, *Nuovo Cimento* **16**, 416 (1960); see also R. Lang, M. O. Scully and W. E. Lamb, Jr., to appear in *Phys. Rev. A* (1973).
6. A. G. Fox and T. Li, *Bell Syst. Tech. J.* **40**, 453 (1961); G. D. Boyd and J. P. Gordon, ibid., p. 489. Cf., also, G. Toraldo di Francia, *QECL* (see Ch. 1, Ref. 1, p. 53.
7. Cf., e.g., D. Bohm, *Quantum Theory*, Prentice-Hall, New York (1951), p. 424.

CHAPTER 6

Semiclassical Theory of the Laser

6.1 INTRODUCTION

DUE TO THE high intensity that may be concentrated in a single mode or in a few modes, the average number of photons per mode in the laser field is extremely high. Thus, for many purposes, it is quite adequate to treat the problem by the semiclassical theory of radiation, in which the atoms are treated by quantum mechanics, but the electromagnetic field is treated in classical terms, i.e., it is not quantized.

We will now describe the semiclassical theory of the laser developed by Lamb[1], as an extension of his treatment of the maser[2]. As our aim is to explain the main physical ideas involved, we will present a simplified version[3] of the theory, keeping the model as simple as possible. Thus, for instance, although the motion of the atoms in the gas laser affects the results significantly, via the Doppler effect, we will neglect it in the treatment, confining ourselves to a qualitative explanation of the way in which the results have to be changed when atomic motion is taken into account.

It is well known that the main defect of the semiclassical radiation theory is that it does not account for spontaneous emission. One of the roles of spontaneous emission in the laser problem is that of a *noise source*. Thus, noise is absent from the semiclassical theory (though it can be artificially introduced), and as a consequence the steady-state solution with only one mode excited is perfectly monochromatic. The intrinsic line width of the laser field, resulting from spontaneous emission and vacuum fluctuations, can only be obtained by taking into account the field quantization. This is done in the quantum theory of the laser, which will be discussed in Section 7. A quantized field theory is also required in order to derive the photon statistics (cf. § 4.2).

Another consequence of the absence of spontaneous emission is that, in the semiclassical theory, we have to assume that there already is a field present in the initial state, for the excited atoms would not decay otherwise. Thus, in order to describe the growth of oscillations from an initial state in which no radiation is present, we also need the quantum theory of the laser.

99

Both in the semiclassical and in the quantum theory of the laser, a number of basic assumptions are made:

(i) TWO-LEVEL ATOMS   We assume that only two atomic levels, corresponding to a and b in Fig. 5.2, take part in the transition that gives rise to the laser radiation. This is justified by the fact that the lasing field modes are very close to resonance with the atomic transition. One can also consider the role played by other levels (cf. Fig. 5.2) in the pumping and losses, and treat three or four-level systems, but it is assumed that only levels a and b are coupled to the laser field.

(ii) ELECTRIC-DIPOLE APPROXIMATION   The interaction between the two-level atoms and the radiation field is treated in the electric-dipole approximation, already described in § 2.2(b). This is again justified by the fact that optical wavelengths are much greater than atomic dimensions. For simplicity it will be assumed that *the electric field in the mode is linearly polarized in the x-direction* (cf. Fig. 5.1), so that the interaction is of the type given by (2.2.9) (actually, as shown in Fig. 5.1, the discharge tube often has Brewster-angle windows, which have ideally 100% transmission for this polarization, but not for the perpendicular one, which is reflected out).

(iii) ABSENCE OF DIRECT INTERATOMIC INTERACTION   It is assumed that any direct interactions among the atoms of the active medium can be neglected: their only coupling arises from the fact that they all interact with a common radiation field. The justification for this assumption lies in the fact that the density of active atoms is that of a highly rarefied gas (cf. § 5.3).

(iv) ROTATING-WAVE APPROXIMATION   In expressions of the type

$$\frac{e^{+i(\omega_n - \omega_0)t - \gamma t}}{\omega_0 - \omega_n - i\gamma} + \frac{e^{-i(\omega_n + \omega_0)t - \gamma t}}{\omega_0 + \omega_n - i\gamma}, \tag{6.1.1}$$

where ω_0 and γ correspond to the atomic frequency and linewidth, and ω_n to the mode frequency, we will always neglect the second (antiresonant) term as compared with the first (resonant) term. This is justified by the fact that the mode is close to resonance, so that $|\omega_0 - \omega_n| \lesssim \gamma$, whereas $\omega_0 + \omega_n \gg \gamma$. The neglected term is, therefore, both small (due to the denominator) and rapidly oscillating. This approximation, which is similar to that already made in going over from (2.2.18) to (2.2.19), is known as the "rotating-wave approximation"[4], for reasons that will become apparent later (cf. § 8.5).

The basic model for the laser is schematized in Fig. 6.1. We have N_0 active 2-level atoms per unit volume, all coupled to the laser field by an electric dipole interaction. To describe the pumping, we can imagine each

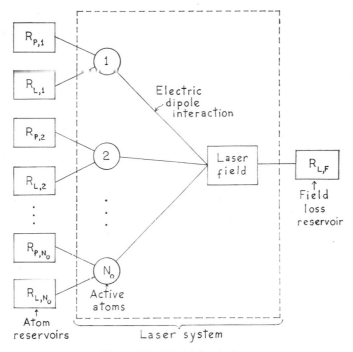

Figure 6.1 Model for the laser

atom i as being coupled to its own "pump reservoir" $R_{P,i}$, which describes, e.g., the energy transfer between He and Ne atoms due to collisions of the 2nd kind; the atoms are pumped independently, i.e., incoherently. Similarly, each atom i is coupled to its own "loss reservoir" $R_{L,i}$, which describes the damping of the levels a and b due, for instance, to (non-laser) transitions to other levels (such as those indicated in Fig. 5.2) and to atomic collisions. The laser field is also coupled to a "field loss reservoir" $R_{L,F}$, which describes the damping of the laser modes due to transmission through the semi-transparent mirror, diffraction losses, losses due to finite conductivity of the walls, etc. It is *assumed* that the exact details of the pumping and loss mechanisms do not play an important role in the theory. Note that the only link between the atoms is through their common interaction with the field, in agreement with assumption (iii).

It should be stressed that the laser system, formed by the active atoms and the field (broken-line box in Fig. 6.1) represents an *open system*, which interchanges energy with the pump and loss reservoirs. In the condition of stable oscillation above threshold, it is a system which is far from thermodynamic equilibrium; this is one of the reasons why it is theoretically interesting; very few models of such systems have been treated so far.

Lamb's semiclassical theory, like Lorentz's classical theory of dielectric media, is a *self-consistent field* theory. As indicated in Fig. 6.2, assuming

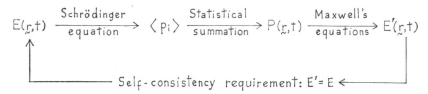

Figure 6.2   Schematic diagram of the self-consistent field method

that we have a field $E(\mathbf{r}, t)$ to begin with, we compute the expectation value of the resulting "microscopic" dipole moment $\langle p_i \rangle$ for a single two-level atom i by means of the Schrödinger equation. We then perform a statistical summation of the microscopic dipole moments over all active atoms, to compute the macroscopic polarization $P(\mathbf{r}, t)$. This now acts as a source term in Maxwell's equations, which allow us to compute the resulting field $E'(\mathbf{r}, t)$. The self-consistency requirement then is that

$$E'(\mathbf{r}, t) = E(\mathbf{r}, t). \tag{6.1.2}$$

This expresses the condition that the induced emission from each atom is due to the average field produced by all the active atoms, to which, in its turn, it contributes.

6.2   THE FIELD EQUATIONS

(a) The frequency and amplitude equations

Let us consider first the equations for the field due to a given macroscopic polarization $\mathbf{P}(\mathbf{r}, t)$. It is customary to write Maxwell's equations in MKS units:

$$\begin{cases} \text{curl } \mathbf{H} = \mathbf{J} + \partial \mathbf{D}/\partial t, \\ \text{curl } \mathbf{E} = -\partial \mathbf{B}/\partial t, \\ \text{div } \mathbf{D} = 0, \\ \text{div } \mathbf{B} = 0, \end{cases} \tag{6.2.1}$$

where

$$\mathbf{D} = \varepsilon_0 \mathbf{E} + \mathbf{P}, \quad \mathbf{B} = \mu_0 \mathbf{H}, \quad \mathbf{J} = \sigma \mathbf{E}. \tag{6.2.2}$$

The "equivalent ohmic conductivity" σ is introduced as a purely phenomenological parameter to represent the damping of the laser field, corresponding to the field loss reservoir $R_{L,F}$ in Fig. 6.1.

Eliminating $\mathbf{E}$ among the equations, we get ($\mu_0 \varepsilon_0 = c^{-2}$)

$$\text{curl curl } \mathbf{E} + \mu_0 \sigma \frac{\partial \mathbf{E}}{\partial t} + \frac{1}{c^2} \frac{\partial^2 \mathbf{E}}{\partial t^2} = -\mu_0 \frac{\partial^2 \mathbf{P}}{\partial t^2}. \qquad (6.2.3)$$

For a gas laser, $\mathbf{P}$ is a small perturbation as compared with $\varepsilon_0 \mathbf{E}$, so that

$$0 = \text{div } \mathbf{D} \approx \varepsilon_0 \text{ div } \mathbf{E}, \qquad (6.2.4)$$

and we can replace curl curl $\mathbf{E}$ by $-\Delta^2\mathbf{E}$; Equ. (6.2.3) then takes the form of the well-known inhomogeneous damped wave equation (telegraph equation).

As explained in § 5.2, we can confine our attention to the axial modes and expand the field in terms of a complete set of such modes, given by (5.2.4):

$$E(z, t) = \sum_n A_n(t) u_n(z) = \sqrt{\frac{2}{L}} \sum_n A_n(t) \sin (k_n z), \qquad (6.2.5)$$

where

$$k_n = n\pi/L = \Omega_n/c, \qquad (6.2.6)$$

Ω_n being the angular frequency associated with the mode. As explained in § 6.1, we have also assumed that the electric field is linearly polarized in the x direction, so that $\mathbf{E} = E\hat{\mathbf{x}}$.

Equation (6.2.3) now becomes

$$\frac{\partial^2 E}{\partial z^2} - \mu_0 \sigma \frac{\partial E}{\partial t} - \frac{1}{c^2} \frac{\partial^2 E}{\partial t^2} = \mu_0 \frac{\partial^2 P}{\partial t^2}. \qquad (6.2.7)$$

Since the solution for the laser field will be practically monochromatic, we can replace

$$\partial^2 P / \partial t^2 \approx -\omega^2 P, \qquad (6.2.8)$$

where ω is the angular frequency of oscillation of P, which will later be identified with the frequency ω_n of the laser field [cf. (6.2.15) and (6.2.18)]. Actually, we have

$$\omega = \omega_n \approx \Omega_n \approx \omega_0, \qquad (6.2.9)$$

where ω_0 is the atomic transition angular frequency.

Substituting (6.2.5) and (6.2.8) in (6.2.7), we find that the unknown mode amplitudes $A_n(t)$ obey the equation of motion for the driven damped harmonic oscillator

$$\ddot{A}_n + \frac{\sigma}{\varepsilon_0} \dot{A}_n + \Omega_n^2 A_n = \frac{\omega^2}{\varepsilon_0} P_n(t), \qquad (6.2.10)$$

where the driving force P_n is the projection of the inhomogeneous source term $P(z, t)$ on the mode n,

$$P_n(t) = \int_0^L P(z, t) u_n(z) \, dz. \qquad (6.2.11)$$

Note that $\sigma/\varepsilon_0 = 1/\tau_n$, where τ_n is the lifetime for the exponential decay of $|A_n|^2$ in the absence of a driving term, according to (6.2.10). Thus, by (5.2.6), we can make the identification [cf., also, (6.2.23)]

$$\sigma/\varepsilon_0 = \Omega_n/Q_n. \tag{6.2.12}$$

Since P_n also depends on the mode amplitudes A_n (according to the loop in Fig. 6.2, they induce the microscopic dipoles whose statistical summation gives rise to P), Equ. (6.2.10), after imposing the self-consistency requirement, will become a differential equation for the determination of A_n. As will be seen later and has already been explained in § 5.3, this is an *essentially nonlinear* differential equation.

The solution of this nonlinear differential equation will be obtained by a classic method due to Krylov and Bogoliubov[5], the "method of slowly-varying amplitudes and phases". One makes the following "Ansatz" for the solution:

$$A_n(t) = E_n(t) \cos [\omega_n t + \varphi_n(t)], \tag{6.2.13}$$

where the "amplitude" $E_n(t)$ and the "phase" $\varphi_n(t)$ are *slowly-varying functions*, which vary little during one period of oscillation $2\pi/\omega_n$, so that, e.g.,

$$\dot{E}_n \ll \omega_n E_n, \quad \dot{\varphi}_n \ll \omega_n \varphi_n, \quad \ddot{E}_n \ll \omega_n \dot{E}_n \ll \omega_n^2 E_n. \tag{6.2.14}$$

The angular frequency ω_n is associated with the oscillations of the field. Due to the quasi-monochromatic character of the field, it is to be expected that the solution will be of this type.

Although (6.2.13) is not a unique representation (a slow variation of $A_n(t)$ can arise either from $E_n(t)$ or from $\varphi_n(t)$), the positive-frequency part of $A_n(t)$ is approximately given by

$$E_n(t) \exp \{-i[\omega_n t + \dot{\varphi}_n(t)]\}, \tag{6.2.15}$$

which represents the associated analytic signal, and, as we have seen in § 1.3, its amplitude and phase are well determined. It will be seen later that, in the rotating-wave approximation, only (6.2.15) gives a significant contribution, so that no ambiguities will arise.

According to (6.2.14), we may neglect the underlined terms in the relations below:

$$
\left.
\begin{aligned}
\dot{A}_n &= -(\omega_n + \dot{\varphi}_n) E_n \sin (\omega_n t + \varphi_n) + \dot{E}_n \cos (\omega_n t + \varphi_n), \\
\ddot{A}_n &= [-(\omega_n^2 + 2\omega_n \dot{\varphi}_n + \dot{\varphi}_n^2) E_n + \underline{\ddot{E}_n}] \cos (\omega_n t + \varphi_n) \\
&\quad - [2(\omega_n + \dot{\varphi}_n) \dot{E}_n + \underline{\ddot{\varphi}_n E_n}] \sin (\omega_n t + \varphi_n).
\end{aligned}
\right\}
\tag{6.2.16}
$$

Let us represent $P_n(t)$ in a similar way. Since P_n will not in general oscillate in phase with A_n, we have to take

$$P_n(t) = C_n(t) \cos [\omega_n t + \varphi_n(t)] + S_n(t) \sin [\omega_n t + \varphi_n(t)], \qquad (6.2.17)$$

where the amplitudes $C_n(t)$ of the "in-phase" component and $S_n(t)$ of the "in-quadrature" component are also slowly-varying functions. Substituting (6.2.15) to (6.2.17) in (6.2.10) and identifying ω with ω_n, we find

$$\left\{ [\Omega_n^2 - (\omega_n^2 + 2\omega_n \dot\varphi_n)] E_n + \frac{\Omega_n}{Q_n} \underline{\dot E_n} \right\} \cos [\omega_n t + \varphi_n(t)]$$

$$+ \left\{ -2\omega_n \dot E_n - \frac{\Omega_n}{Q_n} (\omega_n + \underline{\dot\varphi_n}) E_n \right\} \sin [\omega_n t + \varphi_n(t)]$$

$$= \frac{\omega_n^2}{\varepsilon_0} C_n \cos [\omega_n t + \varphi_n(t)] + \frac{\omega_n^2}{\varepsilon_0} S_n \sin [\omega_n t + \varphi_n(t)], \qquad (6.2.18)$$

where $Q_n \gg 1$ and $\dot\varphi_n \ll \omega_n \varphi_n \sim \omega_n$ are used, in accordance with (6.2.14), to neglect the underlined terms[6]. Also, by (6.2.9), we have

$$\Omega_n^2 - (\omega_n^2 + 2\omega_n \dot\varphi_n) \approx 2\omega_n(\Omega_n - \omega_n - \dot\varphi_n).$$

In order for (6.2.18) to be valid at all times, the coefficients of $\cos [\omega_n t + \varphi_n(t)]$ and $\sin [\omega_n t + \varphi_n(t)]$ must be separately equal, so that we find

$$(\omega_n + \dot\varphi_n - \Omega_n) E_n = -\frac{\omega_n}{2\varepsilon_0} C_n, \qquad (6.2.19)$$

$$\dot E_n + \frac{\Omega_n}{2Q_n} E_n = -\frac{\omega_n}{2\varepsilon_0} S_n. \qquad (6.2.20)$$

Once the driving terms C_n and S_n have been determined from the remainder of the loop in Fig. 6.2, it will be seen that the "in-phase" equation (6.2.19) will allow us to determine the *frequency* of oscillation ω_n, whereas the "in-quadrature" equation (6.2.20) allows us to determine the *amplitude* of oscillation E_n. The phase $\varphi_n(t)$ depends on the initial conditions and it plays practically no role in the solution. The roles of C_n and S_n in determining frequency and amplitude (respectively) become mixed in higher-order approximations of the Krylov-Bogoliubov method, but we shall confine ourselves to the approximation defined by (6.2.19) and (6.2.20), so that we will refer to them as the *frequency equation* and the *amplitude equation*, respectively.

(b) Examples

The role played by the frequency and amplitude equations becomes clear in the following simple special cases.

(I) FREE OSCILLATIONS   In the absence of atomic polarization, $C_n = S_n = 0$, so that the frequency equation yields

$$\omega_n t + \varphi_n = \Omega_n t + \varphi_{n0},$$

i.e., by suitable choice of phase,

$$\omega_n = \Omega_n, \qquad (6.2.21)$$

so that (6.2.19) does determine the frequency. Similarly, (6.2.20) yields

$$E_n = E_{n0} \exp\left(-\frac{\Omega_n t}{2Q_n}\right), \qquad (6.2.22)$$

so that it does determine the amplitude. The free oscillations of the mode thus correspond to the solution

$$A_n(t) = E_{n0} \exp\left(-\frac{\Omega_n t}{2Q_n}\right) \cos(\Omega_n t + \varphi_{n0}), \qquad (6.2.23)$$

confirming the interpretation (6.2.12) of Q_n as the Q-factor associated with the mode n.

(II) LINEAR MEDIUM   A linear medium is one for which (in complex notation)

$$P = \varepsilon_0 \chi E, \qquad (6.2.24)$$

so that it can be described by a *complex susceptibility*

$$\chi = \chi' + i\chi'', \qquad (6.2.25)$$

where, for a gas, $|\chi| \ll 1$, and

$$\eta = \sqrt{\varepsilon/\varepsilon_0} = \sqrt{1 + \chi'} \approx 1 + \frac{\chi'}{2} \qquad (6.2.26)$$

is the corresponding *real refractive index*, and the imaginary part χ'' is associated with the *extinction coefficient* of the medium.

The corresponding *linear approximation* for P_n is

$$P_n = \varepsilon_0 \operatorname{Re}\left[(\chi_n' + i\chi_n'') E_n e^{-i(\omega_n t + \varphi_n)}\right]$$

$$= \varepsilon_0 \chi_n' E_n \cos(\omega_n t + \varphi_n) + \varepsilon_0 \chi_n'' E_n \sin(\omega_n t + \varphi_n). \qquad (6.2.27)$$

Comparing this with (6.2.17), we see that it corresponds to

$$C_n = \varepsilon_0 \chi_n' E_n, \qquad S_n = \varepsilon_0 \chi_n'' E_n, \qquad (6.2.28)$$

and, substituting in the frequency equation (6.2.19), we find

$$\omega_n + \dot{\varphi}_n - \Omega_n = -\frac{\omega_n}{2} \chi_n', \qquad (6.2.29)$$

so that, by suitable choice of phase, as in (6.2.21),

$$\omega_n = \frac{\Omega_n}{1 + (\chi'_n/2)} = \frac{\Omega_n}{\eta_n}, \tag{6.2.30}$$

where η_n is the corresponding real refractive index, given by (6.2.26). This result has a very simple interpretation: the resonant wavelengths [cf. (6.2.6)] $\lambda_n = 2\pi/k_n = 2L/n$ remain unchanged, because they are determined by the geometry, but the corresponding frequencies are now given by $\omega_n = k_n v_n = \Omega_n/\eta_n$, where $v_n = c/\eta_n$ is the phase velocity in the dielectric medium. Note that $\omega_n \approx (1 - (\chi'_n/2))\,\Omega_n$, so that the medium has a "frequency pulling" effect.

The amplitude equation becomes

$$\dot{E}_n + \frac{\Omega_n}{2Q_n} E_n = -\frac{\omega_n}{2} \chi''_n E_n. \tag{6.2.31}$$

Thus, neglecting a small correction,

$$E_n(t) = E_{n0} \exp\left[-\frac{\omega_n}{2}\left(\frac{1}{Q_n} + \chi''_n\right) t \right], \tag{6.2.32}$$

so that, below threshold, in the linear approximation, the atoms give rise to an extinction coefficient associated with χ''_n, as expected.

If we have population inversion and "negative absorption", so that $\chi''_n < 0$, the exponent in (6.2.32) turns positive for

$$-\chi''_n > |\chi''_{n,c}| = 1/Q_n, \tag{6.2.33}$$

where $|\chi''_{n,c}|$ defines the threshold condition [cf. (5.3.17)]. Above the threshold, the amplitude of the laser oscillations would grow exponentially in time, according to (6.2.32). This means that the linear approximation would cease to be valid, and the amplitude of stable oscillation is determined by nonlinear effects.

Therefore, in agreement with the discussion at the end of § 5.3, the linear approximation allows us to determine only the *frequency* (including the linear pulling effect) and the *threshold of oscillation*. In order to obtain the *amplitude of oscillation*, we must take into account nonlinear terms.

6.3   THE MICROSCOPIC POLARIZATION

(a) Schrödinger equation for a 2-level system

According to assumption (i), § 6.1, we may represent an active atom by a two-level system (cf. Fig. 5.2). With the notation defined in Fig. 6.3, and with $\mathscr{H}_A$ denoting the hamiltonian for a free atom, the corresponding normal-

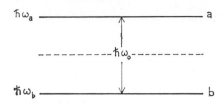

Figure 6.3 Two-level system. The broken line halfway between the levels is taken as the zero of energy in § 6.3(b)

ized stationary eigenfunctions satisfy

$$\mathcal{H}_A\psi_a(\mathbf{x}) = \hbar\omega_a\psi_a(\mathbf{x}); \quad \mathcal{H}_A\psi_b(\mathbf{x}) = \hbar\omega_b\psi_b(\mathbf{x}), \tag{6.3.1}$$

$$(\psi_a, \psi_a) = (\psi_b, \psi_b) = 1; \quad (\psi_a, \psi_b) = 0, \tag{6.3.2}$$

$$\omega_a - \omega_b = \omega_0 \quad \text{(transition frequency)}. \tag{6.3.3}$$

The corresponding time-dependent wave function would be given by (e.g., for a),

$$\psi_a(\mathbf{x}, t) = \psi_a(\mathbf{x}) \exp(-i\omega_a t). \tag{6.3.4}$$

Actually, even in the absence of an external field, the states are not stationary: they decay to other levels by spontaneous emission, and their lifetime is also shortened by atomic collisions (atomic loss reservoirs in Fig. 6.1), as discussed in § 5.3. However, we know that spontaneous emission is not included in semiclassical radiation theory.

In the *Weisskopf-Wigner approximation*[7], the decay from levels a and b takes place exponentially, and (6.3.4) is replaced by

$$\psi(\mathbf{x}, t) = \psi_a(\mathbf{x}) \exp\left[-i\left(\omega_a - i\frac{\gamma_a}{2}\right)t\right], \tag{6.3.5}$$

implying

$$|\psi(\mathbf{x}, t)|^2 = \exp(-\gamma_a t)|\psi(\mathbf{x}, 0)|^2, \tag{6.3.6}$$

so that the corresponding *lifetime* is

$$\tau_a = 1/\gamma_a. \tag{6.3.7}$$

Although (6.3.5) is not valid for extremely short or long times (a Hermitian hamiltonian cannot have complex eigenvalues), it is a very good approximation over the time range of interest. We will, therefore, introduce the atomic loss reservoirs by the phenomenological replacement

$$\omega_a \to \omega_a - i(\gamma_a/2), \quad \omega_b \to \omega_b - i(\gamma_b/2). \tag{6.3.8}$$

The hamiltonian for the atom interacting with the laser field, in the dipole approximation (assumption (ii), § 6.1) is

$$\mathcal{H} = \mathcal{H}_A + \hbar V, \tag{6.3.9}$$

where [cf. (2.2.9)]

$$\hbar V = -ex\, E(\mathbf{r}, t), \tag{6.3.10}$$

where $\mathbf{r}$ is a fixed point in the atom (we have called the interaction $\hbar V$ so that $\hbar$ will drop out of the final equations).

If the atom is pumped to level a at the initial time $t = t_0$, we have

$$\psi(\mathbf{x}, t_0) = \psi_a(\mathbf{x}). \tag{6.3.11}$$

Under the influence of the perturbation $\hbar V$, transitions to level b become possible (induced emission), so that the wave function at a later time t becomes a linear superposition

$$\psi(\mathbf{x}, t) = a(t)\, \psi_a(\mathbf{x}) + b(t)\, \psi_b(\mathbf{x}). \tag{6.3.12}$$

It is convenient to represent this state by the column vector

$$|\psi\rangle_t = \begin{pmatrix} a(t) \\ b(t) \end{pmatrix}, \tag{6.3.13}$$

The Schrödinger equation yields

$$i\hbar \frac{\partial \psi}{\partial t} = i\hbar(\dot{a}\psi_a + \dot{b}\psi_b) = (\mathcal{H}_A + \hbar V)\psi = \hbar[a(\omega_a + V)\psi_a + b(\omega_b + V)\psi_b]. \tag{6.3.14}$$

Taking the scalar products with ψ_a and ψ_b, respectively, with the help of (6.3.2), we find

$$i\dot{a} = \omega_a\, a + V_{ab}\, b,$$
$$i\dot{b} = V_{ba}\, a + \omega_b\, b, \tag{6.3.15}$$

where, by (6.3.10),

$$V_{ab} = (\psi_a, V\psi_b) = V_{ba}^* = -E(\mathbf{r}, t)\, d/\hbar, \tag{6.3.16}$$

and

$$d = e(\psi_a, x\psi_b) = e\langle x_{ab}\rangle = e\int \psi_a^*(\mathbf{x})\, x\psi_b(\mathbf{x})\, d^3x \tag{6.3.17}$$

is the *transition dipole moment* [cf. (2.2.17)]. We can always take the bound-state wave functions ψ_a, ψ_b to be real, so that d and V_{ab} will also be real.

We have also made use of the fact that

$$V_{aa} = V_{bb} = 0, \tag{6.3.18}$$

i.e., the expectation value of the electric dipole moment in a stationary state vanishes. This is because the dipole moment operator has negative parity whereas, due to the definite parity of a stationary state, the probability distribution has positive parity. This is not true, however, for a superposition state such as (6.3.12), provided that a and b have opposite parity, so that the off-diagonal elements V_{ab} do not vanish.

The expectation value $\langle p \rangle$ of the electric dipole moment in the state (6.3.12) is

$$\langle p \rangle = e \langle \psi | x | \psi \rangle = ea^*b(\psi_a, x\psi_b) + eab^*(\psi_b, x\psi_a)$$

$$= d(a^*b + ab^*). \qquad (6.3.19)$$

Let us now introduce the phenomenological damping terms (6.3.8) and rewrite (6.3.15) in matrix notation with the help of (6.3.13):

$$i\frac{d}{dt}\begin{pmatrix} a \\ b \end{pmatrix} = \begin{pmatrix} \omega_a - i(\gamma_a/2) & V_{ab} \\ V_{ab} & \omega_b - i(\gamma_b/2) \end{pmatrix}\begin{pmatrix} a \\ b \end{pmatrix}$$

$$= [H - i(\Gamma/2)]\begin{pmatrix} a \\ b \end{pmatrix}, \qquad (6.3.20)$$

where

$$H = H_0 + V = \begin{pmatrix} \omega_a & 0 \\ 0 & \omega_b \end{pmatrix} + \begin{pmatrix} 0 & V_{ab} \\ V_{ab} & 0 \end{pmatrix} \qquad (6.3.21)$$

and

$$\Gamma = \begin{pmatrix} \gamma_a & 0 \\ 0 & \gamma_b \end{pmatrix}. \qquad (6.3.22)$$

In view of (6.3.19), it is convenient to work with the atomic "density operator" ϱ,

$$\varrho(t) = |\psi\rangle_t \langle\psi|_t = \begin{pmatrix} a \\ b \end{pmatrix}(a^* \; b^*) = \begin{pmatrix} |a|^2 & ab^* \\ a^*b & |b|^2 \end{pmatrix}.$$

If $\Gamma \neq 0$, this is not a true density operator because $\mathrm{Tr}\,\varrho = |a|^2 + |b|^2$ is not equal to 1; even in the absence of V, it decays in time, in accordance with (6.3.5). In terms of ϱ, (6.3.19) becomes

$$\langle p \rangle = d(\varrho_{ab} + \varrho_{ba}) = d(\varrho_{ab} + \varrho_{ab}^*). \qquad (6.3.23)$$

More generally, the expectation value of an observable associated with the operator $\mathcal{O}$ in the state ψ is

$$\langle \mathcal{O} \rangle = \langle \psi | \mathcal{O} | \psi \rangle = \varrho_{aa}\mathcal{O}_{aa} + \varrho_{ab}\mathcal{O}_{ba} + \varrho_{ba}\mathcal{O}_{ab} + \varrho_{bb}\mathcal{O}_{bb} = \mathrm{Tr}\,(\varrho\mathcal{O}),$$

$$(6.3.24)$$

as it would be for a true density operator [cf. (2.1.16)].

An atom in the state a is represented by

$$\varrho(a) = \begin{pmatrix} 1 & 0 \\ 0 & 0 \end{pmatrix}. \qquad (6.3.25)$$

The equation of motion for ϱ follows from (6.3.20) and its hermitian conjugate,

$$i\frac{d\varrho}{dt} - i\binom{a}{b}(a^* b^*) + i\binom{a}{b}(a^* b^*) - \left(H - i\frac{\Gamma}{2}\right)\varrho - \varrho\left(H + i\frac{\Gamma}{2}\right),$$

so that

$$i\frac{d\varrho}{dt} = H\varrho - \varrho H - \frac{i}{2}(\Gamma\varrho + \varrho\Gamma) = [H, \varrho] - \frac{i}{2}\{\Gamma, \varrho\}, \quad (6.3.26)$$

where $\{\Gamma, \varrho\}$ stands for the anticommutator.

(b) Geometrical interpretation

Let us consider the equation of motion (6.3.26) in the special case $\Gamma = 0$, for which ϱ is a true density operator, associated with the pure state $|\psi\rangle_t$.

Any 2×2 matrix can be expanded in the complete set of basis matrices $I, \sigma_1, \sigma_2, \sigma_3$, where I is the identity and the σ's are the Pauli spin matrices. Let

$$\varrho(t) = \frac{1}{2}(\varrho_0 I + \mathbf{r}\cdot\boldsymbol{\sigma}) = \frac{1}{2}\begin{pmatrix} \varrho_0 + r_3 & r_1 - ir_2 \\ r_1 + ir_2 & \varrho_0 - r_3 \end{pmatrix}, \quad (6.3.27)$$

where, by comparison with (6.3.22),

$$\varrho_0 = |a|^2 + |b|^2 = \mathrm{Tr}\,\varrho = 1, \quad (6.3.28)$$

$$r_1 = ab^* + a^*b, \quad (6.3.29)$$

$$r_2 = i(ab^* - a^*b), \quad (6.3.30)$$

$$r_3 = |a|^2 - |b|^2. \quad (6.3.31)$$

Similarly let us decompose H in the same basis, but without necessarily assuming that V_{ab} is real, and let us choose the zero of energy in Fig. 6.3 halfway between the two levels (broken line), so that

$$\omega_a = \omega_0/2 = -\omega_b. \quad (6.3.32)$$

Thus,

$$H = \frac{1}{2}\begin{pmatrix} \omega_0 & V_1 - iV_2 \\ V_1 + iV_2 & -\omega_0 \end{pmatrix} = \frac{1}{2}(V_1\sigma_1 + V_2\sigma_2 + \omega_0\sigma_3), \quad (6.3.33)$$

where we have set

$$V_{ab} = \tfrac{1}{2}(V_1 - iV_2) = V_{ba}^*. \quad (6.3.34)$$

Substituting (6.3.27) and (6.3.33) in the equation of motion

$$i\frac{d\varrho}{dt} = [H, \varrho], \quad (6.3.35)$$

evaluating the commutator with the help of the commutation relations for the Pauli matrices,

$$[\sigma_1, \sigma_2] = 2i\sigma_3 \quad \text{(and circular permutations),} \qquad (6.3.36)$$

and identifying the coefficients of the same basis matrices, we are led to the set of equations

$$\dot{\varrho}_0 = 0, \qquad (6.3.37)$$

$$\dot{r}_1 = V_2 r_3 - \omega_0 r_2, \qquad (6.3.38)$$

$$\dot{r}_2 = \omega_0 r_1 - V_1 r_3, \qquad (6.3.39)$$

$$\dot{r}_3 = V_1 r_2 - V_2 r_1. \qquad (6.3.40)$$

Furthermore, if we express the condition that ϱ describes a pure state (6.3.12) by $\varrho^2 = \varrho$ [cf. (2.1.23)], we find that it leads to

$$r_1^2 + r_2^2 + r_3^2 = \varrho_0^2 = 1. \qquad (6.3.41)$$

The first equation of motion (6.3.37) just expresses the conservation of probability. The remaining three Eqns. (6.3.38)–(6.3.40) take on the remarkably simple form

$$\frac{d\mathbf{r}}{dt} = \boldsymbol{\omega} \times \mathbf{r}, \qquad (6.3.42)$$

if we introduce the vector $\boldsymbol{\omega}$ with components

$$\omega_1 = V_1, \quad \omega_2 = V_2, \quad \omega_3 = \omega_0. \qquad (6.3.43)$$

Furthermore, (6.3.41) becomes

$$\mathbf{r}^2(t) = 1. \qquad (6.3.44)$$

Thus, the time evolution of the state vector for any two-level system under the action of an arbitrary time-dependent perturbation with no diagonal matrix elements (i.e., such that (6.3.18) is valid) can be described very simply in geometrical terms:

We construct a vector $\mathbf{r}(t)$ in a 3-dimensional mathematical space. The components of $\mathbf{r}$ are defined by (6.3.29) to (6.3.31) in terms of the probability amplitudes $a(t)$ and $b(t)$ that define the state of the system according to (6.3.12) (note that a and b are two complex numbers subject to the normalization condition (6.3.28), and the phase of ψ is irrelevant, so that this is a complete characterization). We also construct the 3-vector $\boldsymbol{\omega}$ with components (6.3.43), defined in terms of the perturbation and the circular frequency ω_0 associated with the transition.

According to (6.3.42) and (6.3.44), the time evolution of the two-level system can then be described by saying that *the unit vector $\mathbf{r}(t)$ precesses* *around the vector* $\boldsymbol{\omega}(t)$ *just like a classical gyromagnet precesses in a magnetic field.* In general, both the direction and the magnitude of $\boldsymbol{\omega}$ may change with time.

Note that, by (6.3.27),

so that
$$\langle \sigma_1 \rangle = \text{Tr}\,(\varrho \sigma_1) = \tfrac{1}{2} r_1 \,\text{Tr}\,(\sigma_1^2) = r_1,$$

$$\mathbf{r} = \text{Tr}\,(\varrho \boldsymbol{\sigma}) = \langle \boldsymbol{\sigma} \rangle. \tag{6.3.45}$$

Also, from (6.3.33) and (6.3.43),

$$H = \tfrac{1}{2}\,\boldsymbol{\omega} \cdot \boldsymbol{\sigma}, \tag{6.3.46}$$

so that
$$\langle H \rangle = \text{Tr}\,(\varrho H) = \tfrac{1}{2}\,\boldsymbol{\omega} \cdot \mathbf{r}. \tag{6.3.47}$$

In the particular case in which the 2-level system represents the two magnetic levels of a spin $\tfrac{1}{2}$ particle, $\mathbf{r}$ is proportional to the expectation value of the magnetic moment $\boldsymbol{\mu}$ and $\boldsymbol{\omega}$ is proportional to the magnetic field $\mathbf{H}$, so that in this case the mathematical $\mathbf{r}$ space can be identified with physical space, and the above results describe the familiar precession of a spin $\tfrac{1}{2}$ magnetic moment in a magnetic field. Since the above results are valid for any 2-level system, we see that for any such system we can explore this analogy to make use of the well known techniques introduced by Rabi, Bloch and others in the treatment of magnetic resonance.

In the present case, which corresponds to an electric dipole transition between two levels induced by linearly polarized light, so that $\Delta m = 0$, we have, by (6.3.16) and (6.3.34),

$$V_1 = 2V_{ab} = -2dE(\mathbf{r}, t)/\hbar; \quad V_2 = 0, \tag{6.3.48}$$

and, by (6.3.19) and (6.3.29),

$$\langle p \rangle = dr_1. \tag{6.3.49}$$

The above geometrical analogy, which will be very useful later on, was formulated by Feynman, Vernon and Hellwarth[8].

6.4 THE MACROSCOPIC POLARIZATION

We now come to the next stage of the diagram in Fig. 6.2, the statistical summation over the microscopic dipole moments of all active atoms to get the macroscopic polarization $P(\mathbf{r}, t)$.

A single atom pumped to level a at time t_0 is described at time t by a density operator $\varrho(a, t_0, t)$, which is a solution of the equation of motion (6.3.26) satisfying the initial condition

$$\varrho(a, t_0, t_0) = \varrho(a), \tag{6.4.1}$$

where $\varrho(a)$ is defined by (6.3.25). The corresponding contribution to the expectation value of the dipole moment at time t is given by (6.3.19) and (6.3.23),

$$\langle p(t) \rangle = \text{Tr} \left[\varrho(a, t_0, t) \hat{p} \right] = d[\varrho_{ab}(a, t_0, t) + \varrho_{ab}^*(a, t_0, t)], \tag{6.4.2}$$

where

$$\hat{p} = d\sigma_1 = e \langle x_{ab} \rangle \sigma_1 \tag{6.4.3}$$

is the operator which, according to (6.3.45) and (6.3.49), corresponds to the transition dipole moment. Although we are not explicitly indicating this dependence (to avoid encumbering the notation), ϱ and $\langle p \rangle$ also depend on the atomic position $\mathbf{r}$, because H in (6.3.26) depends on $\mathbf{r}$ through (6.3.16).

Different atoms i are pumped to level a at different times t_{0i}. Thus, the macroscopic polarization is given by

$$P(\mathbf{r}, t) = \sum_{\substack{i \\ t_{0i} < t}} \text{Tr} \left[\varrho(a, t_{0i}, t) \hat{p} \right], \tag{6.4.4}$$

where the sum is extended over all active atoms per unit volume (in a volume element around the point $\mathbf{r}$), which have been pumped to level a up to time t (i.e., for all $t_{0i} < t$).

We can rewrite (6.4.4) as

$$P(\mathbf{r}, t) = \text{Tr} \left[\varrho(a, t) \hat{p} \right], \tag{6.4.5}$$

where

$$\varrho(a, t) = \sum_{\substack{i \\ t_{0i} < t}} \varrho(a, t_{0i}, t). \tag{6.4.6}$$

Let $\lambda_a(t_0) \, dt_0$ be the average number of atoms pumped to level a during $(t_0, t_0 + dt_0)$, per unit volume around $\mathbf{r}$ (we omit the $\mathbf{r}$-dependence in the notation), i.e., the *average pumping rate density*. Then, (6.4.6) may be rewritten as

$$\varrho(a, t) = \int_{-\infty}^{t} \lambda_a(t_0) \, \varrho(a, t_0, t) \, dt_0. \tag{6.4.7}$$

Taking into account (6.4.1) and the equation of motion (6.3.26) obeyed by $\varrho(a, t_0, t)$, we find

$$i \frac{\partial}{\partial t} \varrho(a, t) = i\lambda_a(t) \varrho(a) + \int_{-\infty}^{t} \lambda_a(t_0) \, i \frac{\partial}{\partial t} \varrho(a, t_0, t) \, dt_0$$

$$= i\lambda_a \varrho(a) + [H, \varrho(a, t)] - \frac{i}{2} \{ \Gamma, \varrho(a, t) \}, \tag{6.4.8}$$

where, in the last line, it has been assumed that H *is independent of* t_0, so that it can be removed out of the integral. This assumption is valid in the present treatment, because we are taking the active atoms *at rest* in their positions $\mathbf{r}$, so that the interaction (6.3.16) seen by an atom depends only on its position $\mathbf{r}$ and on t, and not on how long ago it was pumped. When we take account of the motion of the atoms, this is no longer true, because the position of an atom at time t is related to its position at time t_0 by

$$\mathbf{r}(t) = \mathbf{r}(t_0) + \mathbf{v}(t - t_0), \qquad (6.4.9)$$

where $\mathbf{v}$ is the velocity of the atomic motion. Thus, the perturbation seen by a moving atom depends also on t_0, and the simplification corresponding to the last step in (6.4.8) is no longer possible. However, we will ignore this in the present treatment, and postpone to § 6.8 a discussion of the effects of atomic motion (Doppler broadening).

One can also take into account the possibility that there are atoms being excited to the lower level b, e.g., by electronic collisions, and introduce a corresponding rate λ_b. If we then define

$$\varrho(\mathbf{r}, t) = \varrho(a, t) + \varrho(b, t), \qquad (6.4.10)$$

where we have made explicit the $\mathbf{r}$ dependence on the left-hand side, we find for $\varrho(\mathbf{r}, t)$ the equation of motion [cf. (6.4.8)]

$$i\frac{\partial\varrho}{\partial t} = i\lambda + [H, \varrho] - \frac{i}{2}\{\Gamma, \varrho\}, \qquad (6.4.11)$$

where λ is the "pumping rate" matrix

$$\lambda = \lambda_a \varrho(a) + \lambda_b \varrho(b) = \begin{pmatrix} \lambda_a & 0 \\ 0 & \lambda_b \end{pmatrix}. \qquad (6.4.12)$$

The matrix λ is our representation for the atom pump reservoirs in Fig. 6.1, just as Γ represents the atom loss reservoirs. The energy input due to λ can compensate for the energy losses due to Γ and leads to the possibility of a steady-state solution.

The operator (6.4.10) for the active medium, which already contains the statistical summation of Fig. 6.2, is related with the operators $\varrho(a, t_0, t)$ and $\varrho(b, t_0, t)$ associated with single atoms by

$$\varrho(\mathbf{r}, t) = \int\limits_{-\infty}^{t} dt_0 \, [\lambda_a(t_0) \, \varrho(a, t_0, t) + \lambda_b(t_0) \, \varrho(b, t_0, t)]. \qquad (6.4.13)$$

8*

In view of the definitions of λ_a, λ_b and the form (6.3.22) of the single-atom operators, we see that the diagonal elements of $\varrho(\mathbf{r}, t)$ have the following physical interpretation:

$$\varrho_{aa}(\mathbf{r}, t) = N_a(\mathbf{r}, t) = \text{population density of level } a,$$

$$\varrho_{bb}(\mathbf{r}, t) = N_b(\mathbf{r}, t) = \text{population density of level } b,$$

and, by (5.3.13),

$$\varrho_{aa}(\mathbf{r}, t) - \varrho_{bb}(\mathbf{r}, t) = N(\mathbf{r}, t) = \text{density of population inversion.} \quad (6.4.14)$$

Furthermore, according to (6.4.3), (6.4.5), and (6.4.10),

$$P(\mathbf{r}, t) = \text{Tr}[\varrho\,(\mathbf{r}, t)\,\hat{p}] = d[\varrho_{ab}(\mathbf{r}, t) + \varrho_{ab}^*(\mathbf{r}, t)]. \quad (6.4.15)$$

The equation of motion (6.4.11) can also be interpreted[9] in terms of the magnetic resonance analogy described in § 6.3(b); it corresponds to the well known Bloch equations.

We now have the complete set of equations of the self-consistent theory. Equations (6.2.19) and (6.2.20), together with (6.2.5), (6.2.13), (6.2.11) and (6.2.17), determine the electric field $E(z, t)$ due to a given source term $P(z, t)$. On the other hand, the equation of motion (6.4.11), for given pumping and damping rates, together with the hamiltonian (6.3.21), (6.3.16), allows us to determine, via (6.4.15), how the polarization $P(z, t)$ depends on the electric field $E(z, t)$. This dependence is in general nonlinear. The problem is to find a simultaneous solution of these two sets of equations corresponding to steady-state operation.

Let us write down the off-diagonal and diagonal equations of motion for the elements of ϱ corresponding to (6.4.11); with the help of (6.3.21) and (6.3.22), we find

$$i\dot{\varrho}_{ab} = H_{aa}\varrho_{ab} + H_{ab}\varrho_{bb} - \varrho_{aa}H_{ab} - \varrho_{ab}H_{bb} - \frac{i}{2}(\gamma_a\varrho_{ab} + \varrho_{ab}\gamma_b),$$

i.e., with $\omega_0 = \omega_a - \omega_b$, as in (6.3.3), and

$$\gamma_{ab} = \tfrac{1}{2}(\gamma_a + \gamma_b), \quad (6.4.16)$$

we get

$$\dot{\varrho}_{ab} = -i\omega_0\varrho_{ab} - \gamma_{ab}\varrho_{ab} + iV_{ab}(\varrho_{aa} - \varrho_{bb}). \quad (6.4.17)$$

Similarly, for the diagonal equations, we find

$$\dot{\varrho}_{aa} = \lambda_a - \gamma_a\varrho_{aa} + iV_{ab}(\varrho_{ab} - \varrho_{ab}^*), \quad (6.4.18)$$

$$\dot{\varrho}_{bb} = \lambda_b - \gamma_b\varrho_{bb} - iV_{ab}(\varrho_{ab} - \varrho_{ab}^*). \quad (6.4.19)$$

The quantity γ_{ab} defined by (6.4.16) has a simple physical interpretation. The off-diagonal equation of motion for a single atom, given by (6.3.26), has the same form as (6.4.17). Its solution in the absence of an external field, i.e., for $V_{ab} = 0$, is

$$\varrho_{ab}(t) = \varrho_{ab}(0) \exp(-i\omega_0 t - \gamma_{ab}t). \tag{6.4.20}$$

The corresponding atomic transition dipole moment, by (6.3.23), is of the form

$$\langle p \rangle = \beta \cos(\omega_0 t + \varphi_0) \exp(-\gamma_{ab}t). \tag{6.4.21}$$

The atomic line shape factor, as is well known[10], is determined by the Fourier spectrum of (6.4.21), so that it would have a Lorentzian shape, i.e., it would be proportional to

$$\mathcal{L}(\omega - \omega_0) = \frac{1}{(\omega - \omega_0)^2 + \gamma_{ab}^2}. \tag{6.4.22}$$

This is consistent with the fact that, in the present approximation (where the Doppler effect is neglected), the line broadening is due only to transitions to other levels and atomic collisions [cf. (6.3.6)], so that we have the typical lineshape (5.3.8) for homogeneous broadening, with

$$\Delta \nu = \Delta \omega / 2\pi = \gamma_{ab}/\pi. \tag{6.4.23}$$

Taking into account (6.4.16), we see that the width of the atomic line is the sum of the two level widths. This well known result is directly related with the uncertainty principle.

6.5 THE LINEAR APPROXIMATION

The simplest procedure for solving the equations of motion (6.4.17)–(6.4.19) would be an iterative perturbation series in powers of the interaction V_{ab}, the first-order solution being linear in the field $E(z, t)$, and, therefore, corresponding to the linear approximation [cf. (6.4.15)], and the nonlinear effects would be contained in the higher orders of the perturbation series. Although this procedure is followed in the more complicated case in which the Doppler motion is taken into account, it is possible in the present treatment to find a more accurate solution by a different iteration method.

According to (6.4.14), the population inversion density $N(z, t)$ must be derived, in an exact theory, from the simultaneous solution of the equations of motion. However, as explained at the end of § 5.3, we can define the *linear approximation* as that in which *N is assumed to be given and to be time-independent*,

$$N = N(z). \tag{6.5.1}$$

The fact that N is decreased by the operation of the laser (saturation effect) is a nonlinear effect which is thereby neglected. The higher (nonlinear) approximations are then obtained by iteration, starting with the linear approximation.

Substituting (6.4.14) and (6.5.1) in (6.4.17), we get

$$\dot{\varrho}_{ab} + (i\omega_0 + \gamma_{ab}) \varrho_{ab} = iV_{ab}N(z), \tag{6.5.2}$$

where the right-hand side is now linear in the perturbation, i.e., in the field $E(z, t)$ [cf. (6.3.16)], so that, by (6.4.15), we have a linear medium.

Let us assume, to begin with, that only one field mode n is excited in (6.2.5), so that (6.3.16) becomes

$$V_{ab} = -\frac{d}{\hbar} A_n(t) u_n(z) = -\frac{d}{\hbar} E_n(t) u_n(z) \cos{(\omega_n t + \varphi_n)}, \tag{6.5.3}$$

with the help of (6.2.13).

Substituting this in (6.5.2), we get

$$\dot{\varrho}_{ab} + (i\omega_0 + \gamma_{ab}) \varrho_{ab} = -\frac{id}{2\hbar} E_n(t) u_n(z) N(z)$$

$$\times \{\exp{[-i(\omega_n t + \varphi_n)]} + \exp{[i(\omega_n t + \varphi_n)]}\}. \tag{6.5.4}$$

By (6.4.20), we see that the first exponential in (6.5.4) is close to resonance with the atomic transition ($\omega_n \approx \omega_0$), whereas the second exponential ("counter-rotating" wave) is antiresonant. We therefore employ the rotating-wave approximation (assumption (iv), § 6.1), by neglecting the antiresonant term. Setting

$$\varrho_{ab}(t) = \zeta(t) \exp{(-i\omega_n t)},$$

where $\zeta(t)$ is slowly-varying, so that $\dot{\varrho}_{ab} \approx -i\omega_n \varrho_{ab}$, we get the solution

$$\varrho_{ab} = -\frac{id}{2\hbar} \frac{E_n(t) u_n(z)}{[\gamma_{ab} + i(\omega_0 - \omega_n)]} N(z) \exp{[-i(\omega_n t + \varphi_n)]}, \tag{6.5.5}$$

which contains a resonant denominator [cf. (6.1.1)].

Substituting (6.5.5) in (6.4.15) and (6.2.11), we get

$$P_n(t) = \frac{d^2}{2\hbar} E_n(t) \left\{ -i\bar{N}_n \frac{e^{-i(\omega_n t + \varphi_n)}}{[\gamma_{ab} + i(\omega_0 - \omega_n)]} + c.c. \right\}, \tag{6.5.6}$$

where "c.c." denotes the complex conjugate, and

$$\bar{N}_n = \int_0^L N(z, t) u_n^2(z) \, dz, \tag{6.5.7}$$

which, in general, can depend on t, although in the approximation (6.5.1) it does not.

Computing the coefficients of $\cos(\omega_n t + \varphi_n)$ and $\sin(\omega_n t + \varphi_n)$ in (6.5.6) and comparing with (6.2.17), we find, using the notation (6.4.22),

$$C_n = \frac{d^2}{\hbar} \bar{N}_n (\omega_n - \omega_0) \mathcal{L}(\omega_n - \omega_0) E_n, \tag{6.5.8}$$

$$S_n = -\frac{d^2}{\hbar} \bar{N}_n \gamma_{ab} \mathcal{L}(\omega_n - \omega_0) E_n, \tag{6.5.9}$$

which are indeed linear in E_n. Comparing these results with (6.2.25) and (6.2.28), we see that, in this approximation, the active atoms constitute a *linear medium*, with complex susceptibility

$$\chi_n = \frac{d^2}{\varepsilon_0 \hbar} \bar{N}_n \frac{1}{(\omega_n - \omega_0 + i\gamma_{ab})}, \tag{6.5.10}$$

which corresponds to a Lorentz-type dispersion formula.

Substituting (6.5.8) in the amplitude equation (6.2.20), we find, with the help of (6.2.9),

$$\frac{\dot{E}_n}{E_n} = -\frac{\omega_n}{2} \left[\frac{1}{Q_n} - \frac{d^2}{\varepsilon_0 \hbar} \bar{N}_n \gamma_{ab} \mathcal{L}(\omega_n - \omega_0) \right]. \tag{6.5.11}$$

Thus, the *condition for laser oscillation at frequency* ω_n is

$$\frac{d^2}{\varepsilon_0 \hbar} \bar{N}_n \gamma_{ab} \mathcal{L}(\omega_n - \omega_0) \geq \frac{1}{Q_n}, \tag{6.5.12}$$

which can also be obtained directly from (6.5.10) and (6.2.33). Since $\gamma_{ab} \mathcal{L}(\omega_n - \omega_0)$ is proportional to the atomic line shape factor $g(\nu_n)$ [cf. (6.4.23)], this result is the analogue of (5.3.17).

The critical density of population inversion $\bar{N}_{nc}$ is obtained, as in (5.3.18), by taking the maximum $g(\nu_n)$, i.e., by evaluating (6.5.12) at resonance, $\omega_n = \omega_0$. This yields

$$\bar{N}_{n,c} = \frac{\varepsilon_0 \hbar \gamma_{ab}}{d^2 Q_n}. \tag{6.5.13}$$

Noting that $d = e \langle x_{ab} \rangle$ [cf. (6.3.17)] and employing the expression for the fine structure constant in MKS units,

$$\alpha = \frac{e^2}{4\pi \varepsilon_0 \hbar c} \approx \frac{1}{137}, \tag{6.5.14}$$

the above result may be rewritten as

$$\bar{N}_{nc} = \frac{\gamma_{ab}}{4\pi c \alpha \langle x_{ab} \rangle^2 Q_n}. \tag{6.5.15}$$

To compare this with (5.3.18), let us note that the spontaneous emission lifetime τ_{ab} is given by [11]

$$\frac{1}{\tau_{ab}} = \frac{4e^2}{3\hbar}\left(\frac{\omega_0}{c}\right)^3 |\langle \mathbf{r}_{ab}\rangle|^2, \tag{6.5.16}$$

and, by symmetry,

$$\langle x_{ab}\rangle^2 = \langle y_{ab}\rangle^2 = \langle z_{ab}\rangle^2 = \tfrac{1}{3}|\langle \mathbf{r}_{ab}\rangle|^2, \tag{6.5.17}$$

so that we have the correspondence

$$4c\alpha\,\langle x_{ab}\rangle^2 \rightarrow \left(\frac{c}{\omega_0}\right)^3 \frac{1}{\tau_{ab}}. \tag{6.5.18}$$

On the other hand, by (5.2.6), we have, for $\omega_0 = \omega_n$,

$$Q_n = \omega_0\tau_n.$$

Taking into account (6.4.23), we see therefore that (6.5.15) corresponds to

$$\bar{N}_{nc} = \frac{\omega_0^2}{c^3}\,\Delta\nu\,\frac{\tau_{ab}}{\tau_n}. \tag{6.5.19}$$

This is the same as the result (5.3.18) of the elementary theory for Lorentzian line shape [cf. (5.3.19)], except that $\bar{N}_n$ is defined by (6.5.7) more precisely than in the elementary theory, as a sort of average inversion density, taking into account the spatial variation of the mode.

If we now substitute (6.5.8) in the frequency equation (6.2.19), neglecting $\dot{\varphi}_n$, we find

$$\omega_n - \Omega_n = -\frac{d^2\omega_n}{\varepsilon_0\hbar}\,\bar{N}_n(\omega_n - \omega_0)\mathscr{L}(\omega_n - \omega_0). \tag{6.5.20}$$

Taking $\bar{N}_n$ at the value (6.5.12) corresponding to the threshold of oscillation at frequency ω_n, we find

$$\omega_n - \Omega_n = -S(\omega_n - \omega_0), \tag{6.5.21}$$

where

$$S = \frac{\omega_n}{2\gamma_{ab}Q_n}, \tag{6.5.22}$$

In view of (5.2.6) and (6.4.23), this may be rewritten as

$$S = \Delta\Omega_n/\Delta\omega, \tag{6.5.23}$$

i.e., S is the *ratio of the mode width to the atomic linewidth*. The result (6.5.21) represents the *linear pulling effect* already mentioned in connection with (6.2.30).

Solving (6.5.21) with respect to ω_n, we find

$$\omega_n = \frac{\Omega_n + S\omega_0}{1 + S},\qquad(6.5.24)$$

i.e., *the frequency of laser oscillation in mode n is the weighted average of the mode frequency Ω_n and the atomic transition frequency ω_0, with weights 1 and S, respectively.* In particular, at *resonance* ($\Omega_n = \omega_0$),

$$\omega_n = \Omega_n = \omega_0.\qquad(6.5.25)$$

As we see from (6.5.23) and Fig. 5.4, we have $S \ll 1$ for a gas laser, typical values being of order 10^{-1} to 10^{-2}. Thus, by (6.5.23), $\omega_n \approx \Omega_n$, i.e., the laser oscillation frequency is very close to the mode frequency. If we regard the mode and the atom as coupled oscillators, we see that the oscillator with smaller linewidth (larger "Q") pulls the frequency of oscillation closer to its own.

If M modes are excited, oscillating at laser frequencies ω_n ($n = 1, 2, ..., M$), we have to replace (6.5.3) by

$$V_{ab} = -\frac{d}{\hbar} \sum_{n=1}^{M} E_n(t) u_n(z) \cos(\omega_n t + \varphi_n).\qquad(6.5.26)$$

In the linear rotating-wave approximation, the solution corresponding to (6.5.5) then is

$$\varrho_{ab} = -\frac{id}{2\hbar} \sum_{n=1}^{M} \frac{E_n(t) u_n(z)}{[\gamma_{ab} + i(\omega_0 - \omega_n)]} N(z) \exp[-i(\omega_n t + \varphi_n)].\qquad(6.5.27)$$

6.6 NONLINEAR THEORY

(a) Steady-state solution of the atomic equations

In the linear approximation, we took $\varrho_{aa} - \varrho_{bb}$ as a known, time-independent function $N(z)$, and derived the corresponding value (6.5.27) for ϱ_{ab}. The next order approximation, which will already be nonlinear, is obtained by assuming that the population inversion density $\varrho_{aa} - \varrho_{bb} = N(z, t)$ is *slowly-varying in time*, so that at any given time ϱ_{ab} is still given by (6.5.27), but now with $N(z)$ replaced by $N(z, t)$, i.e., for M excited modes,

$$\varrho_{ab} = -\frac{id}{2\hbar}(\varrho_{aa} - \varrho_{bb}) \sum_{n=1}^{M} \frac{E_n(t) u_n(z)}{[\gamma_{ab} + i(\omega_0 - \omega_n)]} \exp[-i(\omega_n t + \varphi_n)].\qquad(6.6.1)$$

One can now substitute (6.6.1) in the diagonal equations (6.4.18), (6.4.19), and solve for $\varrho_{aa} - \varrho_{bb}$.

By (6.5.26) and (6.6.1),

$$iV_{ab}(\varrho_{ab} - \varrho_{ab}^*) = -\frac{d^2}{2\hbar^2} \sum_{m=1}^{M} E_m u_m \cos(\omega_m t + \varphi_m)(\varrho_{aa} - \varrho_{bb})$$

$$\times \sum_{n=1}^{M} \left\{ \frac{E_n u_n}{[\gamma_{ab} + i(\omega_0 - \omega_n)]} \exp[-i(\omega_n t + \varphi_n)] + c.c. \right\}.$$

Discarding antiresonant terms (rotating-wave approximation), this leads to

$$iV_{ab}(\varrho_{ab} - \varrho_{ab}^*) = -R(\varrho_{aa} - \varrho_{bb}), \tag{6.6.2}$$

where

$$R = \frac{d^2}{4\hbar^2} \sum_{m=1}^{M} \sum_{n=1}^{M} \left\{ \frac{E_m E_n u_m u_n}{[\gamma_{ab} + i(\omega_0 - \omega_n)]} \exp\{i[(\omega_m - \omega_n)t + \varphi_m - \varphi_n]\} + c.c. \right\}. \tag{6.6.3}$$

Substituting (6.6.2) in (6.4.18), (6.4.19), we get

$$\dot{\varrho}_{aa} = \lambda_a - \gamma_a \varrho_{aa} + R(\varrho_{bb} - \varrho_{aa}), \tag{6.6.4}$$

$$\dot{\varrho}_{bb} = \lambda_b - \gamma_b \varrho_{bb} + R(\varrho_{aa} - \varrho_{bb}). \tag{6.6.5}$$

These equations have the form of *rate equations* for the population densities ϱ_{aa} and ϱ_{bb} of levels a and b. The first two terms represent the pumping and damping rates associated with the atom reservoirs in Fig. 6.1. The last term represents the effects of induced emission and absorption, which decrease (increase) the population of level a (b) when there is population inversion ($\varrho_{aa} - \varrho_{bb} > 0$). In fact, as will be shown below, $R \geq 0$ and we can interpret R as an *induced transition rate*. Note that R is quadratic in the mode amplitudes, so that it is proportional to the intensity.

If more than one mode is excited, we see from (6.6.3) that R contains pulsating components, corresponding to frequencies $\omega_m - \omega_n$. They would lead to pulsation effects also in the population densities, and the polarization, via (6.4.15) and (6.6.1), would contain combination tones, as is characteristic of nonlinear oscillations. However, for a typical gas laser, $\omega_m - \omega_n \gg \gamma_{ab}$ (the separation between adjacent axial modes is of the order of 150 Mc, whereas $\gamma_{ab} \sim 10$ Mc), so that pulsating components are relatively small and we will neglect them here. Thus, we restrict the summation in (6.6.3) to $m = n$, which leads to

$$R = \frac{d^2}{2\hbar^2} \sum_{n=1}^{M} \gamma_{ab} \mathscr{L}(\omega_n - \omega_0) E_n^2(t) u_n^2(z) \geq 0. \tag{6.6.6}$$

If we consider a single mode, $\gamma_{ab}\mathscr{L}(\omega_n - \omega_0)$ is proportional to the atomic line shape factor $g(\nu_n)$, $E_n^2 u_n^2$ to the intensity $I(\nu_n)$ and $d^2/\hbar^2$ to the inverse of the spontaneous emission lifetime τ_{ab} [cf. (6.5.18)], so that (6.6.6) indeed corresponds to the induced transition rate (5.3.7).

The *steady-state solution* of (6.6.4), (6.6.5) is obtained by setting

$$\dot{\varrho}_{aa} = \dot{\varrho}_{bb} = 0, \tag{6.6.7}$$

which leads to

$$N = \varrho_{aa} - \varrho_{bb} = \frac{N^{(0)}}{1 + (R/R_s),} \tag{6.6.8}$$

where

$$N^{(0)} = \frac{\lambda_a}{\gamma_a} - \frac{\lambda_b}{\gamma_b}, \tag{6.6.9}$$

and [cf. (6.4.16), (6.3.7)]

$$\frac{1}{R_s} = \frac{1}{\gamma_a} + \frac{1}{\gamma_b} = \frac{2\gamma_{ab}}{\gamma_a \gamma_b} = \tau_a + \tau_b. \tag{6.6.10}$$

In the absence of a radiation field (i.e., for $R = 0$), (6.6.8) yields $N = N^{(0)}$. Thus, $N^{(0)}$ represents the *zero-field inversion density*, resulting from equilibrium between pumping and loss rates. In the presence of a field, the inversion density is reduced by a factor $1 + (R/R_s)$, as shown by (6.6.8). This represents the effect of induced emission; the reduction increases with the intensity, as does the induced transition rate R. This is the nonlinear *saturation effect* referred to at the end of § 5.3, which prevents the exponential growth of the intensity found in the linear approximation and determines the stable amplitude of oscillation of the laser.

Note that N falls to half its zero-field value for $R = R_s$, so that R_s may be thought of as a *saturation rate*. By (6.6.10), it is the inverse of the total lifetime associated with emission from the two levels. Since R is a slowly-varying function of t and z, according to (6.6.6), so is $N(z, t)$ given by (6.6.8), which is consistent with the initial assumption that led to the approximation (6.6.1).

Substituting (6.6.8) in (6.6.1), we find the value of ϱ_{ab} in this nonlinear approximation:

$$\varrho_{ab} = -\frac{id}{2\hbar} \frac{N^{(0)}}{[1 + (R/R_s)]} \sum_{n=1}^{M} \frac{E_n(t) u_n(z)}{[\gamma_{ab} + i(\omega_0 - \omega_n)]} \exp[-i(\omega_n t + \varphi_n)]. \tag{6.6.11}$$

This differs from the linear approximation (6.5.27) only by the substitution

$$N(z) \rightarrow \frac{N^{(0)}}{1 + (R/R_s)}. \tag{6.6.12}$$

(b) The field equations

Having found the solution (6.6.8), (6.6.11) of the atomic equations, we must now substitute (6.6.11) in (6.4.15) and (6.2.11) in order to find the corresponding values of C_n and S_n. According to (6.6.12), they differ from the corresponding values (6.5.8), (6.5.9) in the linear approximation only by the substitution [cf. (6.5.7)]

$$\bar{N}_n \to \bar{N}'_n = \int_0^L \frac{N^{(0)}}{1 + (R/R_s)} u_n^2(z) \, dz. \tag{6.6.13}$$

By (6.6.6), the denominator is now a quadratic function of the field amplitudes E_n, so that the amplitude and phase equations (6.2.19), (6.2.20) become a complicated system of simultaneous nonlinear equations.

Let us consider first the case in which there is only one excited mode n, with sufficiently weak intensity that R is well below the saturation rate, i.e.,

$$R/R_s \ll 1. \tag{6.6.14}$$

We may then employ the expansion

$$\frac{1}{1 + (R/R_s)} = 1 - \frac{R}{R_s} + \cdots \tag{6.6.15}$$

and stop at the second term. Substituting this in (6.6.13) and taking into account (6.6.6) and (6.6.10), we find

$$\bar{N}'_n = \bar{N} - \frac{d^2 \gamma_{ab}^2}{\hbar^2 \gamma_a \gamma_b} \mathscr{L}(\omega_n - \omega_0) \int N^{(0)} u_n^4(z) \, dz \, E_n^2, \tag{6.6.16}$$

so that this corresponds to keeping just the quadratic correction in the field amplitude. We have introduced

$$\bar{N} = \int_0^L N^{(0)} u_n^2(z) \, dz. \tag{6.6.17}$$

The pumping rates in (6.6.9) may depend on z; usually, however, it can be assumed that $N^{(0)}$ is either constant or slowly varying, i.e., it changes little within a wavelength. Taking into account (5.2.4), it then follows that

$$\frac{1}{\bar{N}} \int_0^L N^{(0)} u_n^4(z) \, dz \approx \frac{2}{L} \frac{\int_0^{n\pi} \sin^4 x \, dx}{\int_0^{n\pi} \sin^2 x \, dx} = \frac{3}{2L}, \tag{6.6.18}$$

so that (6.6.16) becomes

$$\bar{N}'_n = \bar{N}[1 - \gamma_{ab}^2 \mathscr{L}(\omega_n - \omega_0) I_n], \tag{6.6.19}$$

where

$$I_n = \frac{3}{2L} \frac{d^2}{\hbar^2 \gamma_a \gamma_b} E_n^2 \qquad (6.6.20)$$

is a dimensionless parameter that measures the *intensity* of mode n.

Within the approximation (6.6.15), we may also rewrite (6.6.19) as

$$\bar{N}_n' \approx \frac{\bar{N}}{1 + \gamma_{ab}^2 \mathscr{L}(\omega_n - \omega_0) I_n} = \frac{\bar{N}\mathscr{L}^{-1}(\omega_n - \omega_0)}{(\omega_n - \omega_0)^2 + (1 + I_n)\gamma_{ab}^2}, \qquad (6.6.21)$$

Substituting (6.6.13) and (6.6.21) in (6.5.9), we find

$$S_n = -\frac{d^2}{\hbar} \bar{N} E_n \frac{\gamma_{ab}}{(\omega_n - \omega_0)^2 + (1 + I_n)\gamma_{ab}^2}. \qquad (6.6.22)$$

Comparing this with (6.5.9), we see that the effective resonance width has increased by a factor $\sqrt{1 + I_n}$ due to induced emission.

Substituting (6.6.22) in the amplitude equation (6.2.20), we see that (6.5.11) is replaced by (taking $\Omega_n \approx \omega_n$)

$$-\frac{2}{\omega_n} \frac{\dot{E}_n}{E_n} = \frac{1}{Q_n} - \frac{d^2}{\varepsilon_0 \hbar} \bar{N} \frac{\gamma_{ab}}{(\omega_n - \omega_0)^2 + \gamma_{ab}^2(1 + I_n)}. \qquad (6.6.23)$$

This nonlinear differential equation now has a *steady-state solution*, $\dot{E}_n = 0$, with E_n (or, equivalently, I_n) given by setting the right-hand side equal to zero, i.e.,

$$(\omega_n - \omega_0)^2 + \gamma_{ab}^2(1 + I_n) = \frac{d^2 Q_n}{\varepsilon_0 \hbar \gamma_{ab}} \bar{N}\gamma_{ab}^2 = \frac{\bar{N}}{\bar{N}_c}\gamma_{ab}^2,$$

where $\bar{N}_c$ is the critical inversion density (6.5.13) associated with the threshold. Solving with respect to I_n, we find

$$I_n = \frac{\bar{N}}{\bar{N}_c} - 1 - \frac{(\omega_n - \omega_0)^2}{\gamma_{ab}^2}. \qquad (6.6.24)$$

In particular, at resonance ($\omega_n = \omega_0$),

$$I_n = \frac{\bar{N}}{\bar{N}_c} - 1 \quad (\omega_n = \omega_0), \qquad (6.6.25)$$

i.e., *the dimensionless intensity of mode n is equal to the fractional excess of the inversion density over the critical inversion*. In particular, for $\bar{N} = \bar{N}_c$, we have $I_n = 0$, in agreement with the definition of the threshold. The ratio $\bar{N}/\bar{N}_c$ is called the *relative excitation*. Outside of resonance, according to (6.6.24), the intensity decreases by the square of the ratio of the detuning $\omega_n - \omega_0$ to the linewidth γ_{ab}.

According to (6.6.19), the condition (6.6.14) for the validity of the above results is that

$$I_n \ll 1 + \frac{(\omega_n - \omega_0)^2}{\gamma_{ab}^2}, \tag{6.6.26}$$

so that the results do not apply to a laser oscillating high above threshold.

6.7 MODE COMPETITION

Let us now see what happens when we consider the possibility that more than one mode may be excited, still within the approximation (6.6.14). Substituting (6.6.6) and (6.6.15) in (6.6.13), we now get

$$\bar{N}_n' = \bar{N} - \frac{3}{2L} \frac{d^2}{\hbar^2 \gamma_a \gamma_b} \gamma_{ab}^2 \mathscr{L}(\omega_n - \omega_0) \, \bar{N} E_n^2$$

$$- \frac{d^2}{\hbar^2 \gamma_a \gamma_b} \gamma_{ab}^2 \sum_{m \neq n} \mathscr{L}(\omega_m - \omega_0) \int_0^L N^{(0)} u_m^2(z) u_n^2(z) dz \, E_m^2, \tag{6.7.1}$$

where the term $m = n$ has been replaced by its expression (6.6.19)–(6.6.20). By (6.2.5), the last integral in (6.7.1) is given by

$$\frac{4}{L^2} \int_0^L N^{(0)} \sin^2(k_m z) \sin^2(k_n z) \, dz = \frac{1}{L^2} \int_0^L N^{(0)}\{1 - \cos(2k_m z)$$

$$- \cos(2k_n z) + \tfrac{1}{2} \cos[2(k_m + k_n) z] + \tfrac{1}{2} \cos[2(k_m - k_n) z]\} \, dz.$$

If $N^{(0)}$ is either constant or slowly varying, as was already assumed in connection with (6.6.17), only the first and last terms within curly brackets give an appreciable contribution. Let us define

$$N_{2(m-n)} = \frac{1}{L} \int_0^L N^{(0)} \cos[2(k_m - k_n)z] \, dz, \tag{6.7.2}$$

and note that, for $m = n$ [cf. (6.6.17)],

$$N_0 = \frac{1}{L} \int_0^L N^{(0)} \, dz \approx \frac{2}{L} \int_0^L N^{(0)} \sin^2(k_n z) \, dz = \bar{N}. \tag{6.7.3}$$

Then,

$$\int_0^L N^{(0)} u_m^2(z) \, u_n^2(z) \, dz \approx \frac{1}{L} (\bar{N} + \tfrac{1}{2} N_{2(m-n)}). \tag{6.7.4}$$

Substituting (6.6.13), (6.7.1) and (6.7.4) in (6.5.9), and substituting the result in the amplitude equation (6.2.20), we get the following nonlinear system of coupled differential equations:

$$\dot{E}_n = \alpha_n E_n - \beta_n E_n^3 - \sum_{\substack{m \neq n \\ 1}}^{M} \theta_{mn} E_n E_m^2, \qquad n = 1, 2, \ldots, M, \qquad (6.7.5)$$

where

$$\alpha_n = \frac{\omega_n d^2}{2\varepsilon_0 \hbar \gamma_{ab}} \, \bar{N}\gamma_{ab}^2 \, \mathscr{L}(\omega_n - \omega_0) - \frac{\Omega_n}{2Q_n}, \qquad (6.7.6)$$

$$\beta_n = \frac{\omega_n}{2\varepsilon_0} \frac{d^4 \gamma_{ab}^3}{\hbar^3 \gamma_a \gamma_b} \, \mathscr{L}^2(\omega_n - \omega_0) \cdot \frac{3}{2} \frac{\bar{N}}{L}, \qquad (6.7.7)$$

$$\theta_{mn} = \frac{\omega_n}{2\varepsilon_0} \frac{d^4 \gamma_{ab}^3}{\hbar^3 \gamma_a \gamma_b} \, \mathscr{L}(\omega_m - \omega_0)\mathscr{L}(\omega_n - \omega_0)\frac{1}{L}\left(\bar{N} + \tfrac{1}{2} N_{2(m-n)}\right) \approx \theta_{nm}. \qquad (6.7.8)$$

The parameter α_n, as we see by reference to (6.5.11), represents the *overall gain*: the condition for laser oscillation in the mode n is $\alpha_n \geqq 0$ [cf. (6.5.12), (5.3.17)]. The first term in α_n represents the gain due to the balance between the atom pump and loss reservoirs [which defines $\bar{N}$, by (6.6.9) and (6.6.17)] and the amplifying effect of induced transitions. The second term in α_n represents the loss due to the field loss reservoir. Thus,

$$\alpha_n = \text{gain (atom reservoirs + induced transitions)}$$

$$- \text{loss (field reservoir)} \qquad (6.7.9)$$

The parameter β_n is a *saturation parameter*. To see this, consider (6.7.5) when a single mode is excited:

$$\dot{E}_n = \alpha_n E_n - \beta_n E_n^3. \qquad (6.7.10)$$

It is readily verified that this coincides with (6.6.23) within the present approximation [it suffices to replace $\bar{N}_n$ by (6.6.19) in (6.5.11)]. The term in E_n^3 corresponds to $E_n I_n$, which represents, as we have seen, a nonlinear saturation effect. Note that $\beta_n > 0$ for positive population inversion ($\bar{N} > 0$).

From (6.7.10) we can recover the analogue of (6.6.24), i.e., the intensity of single-mode laser oscillation in the steady state ($\dot{E}_n = 0$), in the suggestive form

$$E_n^2 = \frac{\alpha_n}{\beta_n} = \frac{\text{gain} - \text{loss}}{\text{saturation parameter}}. \qquad (6.7.11)$$

The parameters θ_{mn} in (6.7.8) represent the nonlinear saturation effect on the *coupling between different modes*. Note that they depend on the overlap between the corresponding Lorentzian factors. Let us illustrate the effect of competition among modes by considering the simplest case, in which there are only two adjacent modes (denoted by 1 and 2) above threshold. Equations (6.7.5) become

$$\dot{E}_1 = \alpha_1 E_1 - \beta_1 E_1^3 - \theta E_1 E_2^2, \quad \dot{E}_2 = \alpha_2 E_2 - \beta_2 E_2^3 - \theta E_2 E_1^2,$$

(6.7.12)

where [cf. (6.7.8)]

$$\theta = \theta_{12} \approx \theta_{21} > 0, \quad \alpha_i > 0, \quad \beta_i > 0 \quad (i = 1, 2).$$ (6.7.13)

With the substitution

$$E_1^2 = x, \quad E_2^2 = y,$$ (6.7.14)

equations (6.7.12) become

$$\tfrac{1}{2}\dot{x} = (\alpha_1 - \beta_1 x - \theta y)\, x, \quad \tfrac{1}{2}\dot{y} = (\alpha_2 - \beta_2 y - \theta x)\, y. \quad (6.7.15)$$

The condition for a steady state is $\dot{x} = \dot{y} = 0$. According to (6.7.14), only solutions in the first quadrant ($x \geq 0$, $y \geq 0$) make sense. We see from (6.7.15) that only four types of stationary solutions are possible:

(I) $\bar{y} = 0, \quad \bar{x} = \alpha_1/\beta_1.$ (6.7.16)

(II) $\bar{x} = 0, \quad \bar{y} = \alpha_2/\beta_2.$ (6.7.17)

(III) $(\bar{x}, \bar{y}) =$ simultaneous solution of $\begin{cases}(L_1)\ \beta_1 x + \theta y = \alpha_1 \\ (L_2)\ \theta x + \beta_2 y = \alpha_2\end{cases} (x > 0, y > 0).$

(6.7.18)

(IV) $\bar{x} = \bar{y} = 0.$ (6.7.19)

Solution (I) would represent laser oscillation in mode 1 alone, and (II) in mode 2 alone [cf. (6.7.11)]. Solution (III), if it exists, i.e., if the straight lines L_1 and L_2 intersect in the first quadrant, represents simultaneous oscillation in both modes. Finally, (IV) is the trivial solution, which we expect to be unstable above threshold.

In order to investigate which of the possible stationary solutions represent stable equilibrium states and which ones are unstable, one can employ the method of the "phase portrait", due to Poincaré, which is frequently employed to investigate the stability of solutions of nonlinear differential equations.

To each solution $x(t)$, $y(t)$ of (6.7.15) at time t we associate a point with coordinates (x, y) in the "phase plane". For each set of initial conditions $x_0 = x(t_0)$, $y_0 = y(t_0)$, we obtain a "phase curve" $y(x)$ (with parametric equations $x = x(t)$, $y = y(t)$) going through (x_0, y_0) and representing the evolution of the system beyond this point. A stationary solution $(\bar{x}, \bar{y})$ is *stable* if all phase curves in the neighborhood of $(\bar{x}, \bar{y})$ tend towards this point; it is *unstable* if they tend away from it.

According to (6.7.15), the slope of the tangent to the phase curve $y(x)$ at the point (x, y) is given by

$$\frac{dy}{dx} = \frac{\dot{y}}{\dot{x}} = \frac{y}{x} \frac{(\alpha_2 - \theta x - \beta_2 y)}{(\alpha_1 - \beta_1 x - \theta y)}, \tag{6.7.20}$$

Thus, all phase curves that cross L_1 (L_2) have vertical (horizontal) tangent at the crossing point. The denominator (numerator) of (6.7.20) is positive below the straight line L_1 (L_2), which passes through the point (I) [(II)] and has slope $-\beta_1/\theta$ $(-\theta/\beta_2)$. These results suffice for a qualitative discussion of the stability problem.

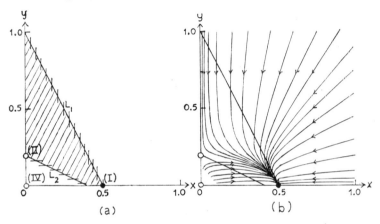

(a)   (b)

Figure 6.4 Phase curves for mode 1 well above threshold ($\alpha_1 = 1$) and mode 2 having smaller gain ($\alpha_2 = 0.4$); $\beta_1 = \beta_2 = 2$; $\theta = 1$. The slope of the phase curves is infinite where they cross L_1, zero where they cross L_2, and negative in the shaded region. ○ – unstable solution; ● – stable solution. Only mode 1 oscillates, quenching oscillation in mode 2 (after W. E. Lamb, *Phys. Rev.* **134**, A 1429 (1964)).

Let us consider first the case in which mode 1 is well above threshold, but mode 2 has appreciably smaller gain. A typical situation is illustrated in Fig. 6.4. Figure 6.4(a) corresponds to the qualitative discussion explained above. The phase curves have negative slope in the shaded region and positive slope elsewhere. Solution (III) does not exist in this case, and (IV) is always unstable. Thus, according to Fig. 6.4(a), (II) must be unstable and (I) must be stable. This is confirmed by a computer integration, which leads to the phase curves shown in Fig. 6.4(b). Whatever the starting point, the system always evolves towards the stable operating point (I). Oscillation in mode 2 is *quenched* by the oscillation in mode 1: the "effective gain" for

mode 2, given by $\alpha_2' = \alpha_2 - \theta\bar{x} = \alpha_2 - \theta\dfrac{\alpha_1}{\beta_1}$, is made negative by oscillation in mode 1 (in the example of Fig. 6.4, $\alpha_2' = -0.1$).

If the gain of mode 2 is increased until the effective gain becomes positive, there are two possibilities, depending on whether the mode coupling, measured by θ, is weak or strong. This classification depends on the relation between the slope $-\beta_1/\theta$ of L_1 and the slope $-\theta/\beta_2$ of L_2:

$$\theta^2 < \beta_1\beta_2 \text{ (weak coupling)}; \quad \theta^2 > \beta_1\beta_2 \text{ (strong coupling)}. \quad (6.7.21)$$

A typical case of *weak coupling* is illustrated in Fig. 6.5 (Fig. 6.5(a) represents the qualitative treatment). All phase curves lead to point (III), so that the stable operating point corresponds to *simultaneous oscillation in both modes*, with amplitudes corresponding to the solution (III) of (6.7.18).

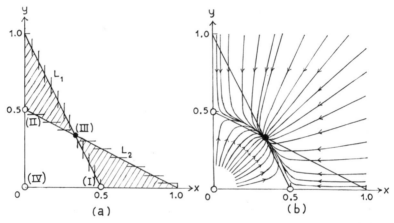

Figure 6.5 Phase curves for both modes well above threshold ($\alpha_1 = \alpha_2 = 1$) and weak coupling ($\beta_1 = \beta_2 = 2$; $\theta = 1$). The stable solution (*III*) corresponds to simultaneous oscillation in both modes. The conventions are the same as in Fig. 6.4 [after W. E. Lamb, *Phys. Rev.* **134**, A 1429 (1964)]

On the other hand, for *strong coupling* (Fig. 6.6), point (III) corresponds to unstable equilibrium. The system has two stable equilibrium positions, corresponding to oscillation in mode 1 alone (I) or in mode 2 alone (II). Which of these is chosen depends on the initial conditions, i.e., on the past history of the system; this leads to *hysteresis* effects. In the symmetric case illustrated in Fig. 6.6, the system approaches mode 1 or 2 depending on whether the initial position is below or above the first bisector, respectively, i.e., the initially "richer" mode wins the competition. This is related with the quenching effect discussed above.

In the limiting case $\beta_1/\theta = \theta/\beta_2$, the straight lines L_1 and L_2 coincide, and any point on the resulting straight line is a neutral equilibrium position, but the system in this case would be very sensitive to perturbations.

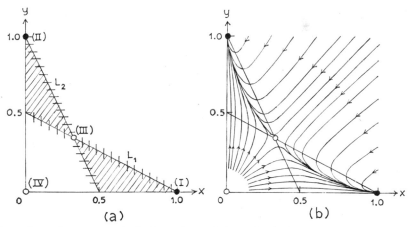

Figure 6.6 Phase curves for both modes well above threshold ($\alpha_1 = \alpha_2 = 1$) and strong coupling ($\beta_1 = \beta_2 = 1$; $\theta = 2$). There are two stable solutions, corresponding to oscillation either in mode 1 (*I*) or 2 (*II*). The operating point depends on the initial conditions; hysteresis occurs. The conventions are the same as in Fig. 6.4 [after W. E. Lamb, *Phys. Rev.* **134**, A 1429 (1964)].

In a gas laser conditions are usually such that one has weak coupling, so that one ordinarily has simultaneous oscillation in various modes.

The above treatment of the interaction between nonlinear oscillations is closely related with Van der Pol's theory[12] of nonlinear oscillations in a triode oscillator. Van der Pol took as the operating point in the current × voltage characteristic for the triode an inflection point, approximating it by a cubic,

$$i = -aV + bV^3. \tag{6.7.22}$$

This may be compared with (6.7.10). Besides considering the usual oscillator with only one resonant frequency, he also treated the circuit shown in Fig. 6.7, which has two resonant frequencies, ω_1 und ω_2. Taking

$$V = E_1 \cos(\omega_1 t) + E_2 \cos(\omega_2 t), \tag{6.7.23}$$

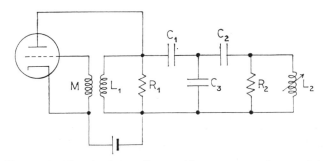

Figure 6.7 Van der Pol oscillator with two resonant frequencies.

substituting in (6.7.22), and retaining only terms of angular frequency ω_1 or ω_2, one finds that the result corresponds to (6.7.12) with $\theta = 2\beta_1 = 2\beta_2$, i.e., to strong coupling, so that the circuit oscillates either in mode 1 or in mode 2. Similar nonlinear methods have also been employed to treat biological competition among different species.[13]

There remains to discuss the frequency equation in the nonlinear approximation. We see from (6.5.8) and (6.5.9) that the value of C_n corresponding to a given approximation for S_n is

$$C_n = (\omega_0 - \omega_n) S_n / \gamma_{ab}. \tag{6.7.24}$$

Substituting $\bar{N}_n$ by (6.6.19) in (6.5.8) and substituting the result in the frequency equation (6.2.19), one would find that (6.5.24) is replaced by a result of the form

$$\omega_n = \Omega_n + \sigma_n + \varrho_n E_n^2, \tag{6.7.25}$$

so that, in addition to the linear pulling effect described in § 6.5, we also have a "power pushing" effect: the frequency depends on the amplitude of oscillation, as is characteristic for an anharmonic oscillator. Similar effects occur when several modes are excited. Note that *the laser field in single-mode oscillation is perfectly monochromatic in the semiclassical theory*, as was remarked in § 6.1.

6.8   EFFECT OF ATOMIC MOTION

Thus far we have treated the active atoms as if they were at rest. Actually, in a typical gas laser, the mean thermal velocity is sufficiently large for the atom to move through many optical wavelengths before decaying, so that the frequencies "seen" by the atom are changed by the Doppler effect. This leads to inhomogeneous broadening of the atomic line (§ 5.3) and introduces important corrections.

The atoms pumped to level a or b at time t_0, in the neighborhood of z_0, are located, at a later time t, around

$$z = z_0 + v(t - t_0), \tag{6.8.1}$$

where v is the z-component of the atomic velocity. Thus, instead of (6.5.26), this group of atoms "sees" the perturbation

$$V_{ab}(z_0, t_0, t, v) = -\frac{d}{\hbar} \sum_{n=1}^{M} E_n u_n[z_0 + v(t - t_0)] \cos(\omega_n t + \varphi_n). \tag{6.8.2}$$

Substituting u_n by its expression (6.2.5), it follows that the oscillation frequencies, as seen by the atoms, are Doppler-shifted from ω_n to

$$\omega_n \pm k_n v. \tag{6.8.3}$$

As explained in § 6.4, due to the explicit dependence on t_0 in (6.8.2), the equations of motion can no longer be written in the form of the last line of (6.4.8). One has to go back to the microscopic equation of motion (6.3.26), to solve it for each group of atoms, obtaining $\varrho(a, z_0, t_0, t, v)$, and then one can derive $\varrho(a, t)$ by summing over (z_0, v), in addition to the sum over t_0. This procedure was applied by Lamb[14], who solved (6.3.26) by standard time-dependent perturbation theory. In this method, the contribution to ϱ_{ab} due to first-order perturbation theory is linear in the field [cf. (6.8.2)], so that it corresponds to the linear approximation. The second-order term vanishes and the third-order contribution is cubic in the field, corresponding to the nonlinear approximation of the previous treatment.

One can also derive some of the results by making a more drastic approximation, in which one neglects the dependence of (6.8.2) on z_0 and t_0, retaining only the dependence on v. In this case, one can apply the previous treatment for each v and then integrate over the velocity distribution. In view of (6.8.3), one finds that (6.6.6) is replaced by

$$R(z,v) = \frac{d^2}{4\hbar^2} \sum_{n=1}^{M} \gamma_{ab}[\mathscr{L}(\omega_n - \omega_0 + k_n v) + \mathscr{L}(\omega_n - \omega_0 - k_n v)]E_n^2(t)u_n^2(z).$$

(6.8.4)

The pumping rates now have to be defined in terms of the average number of excited atoms with velocity v, e.g.,

$$\lambda_a(z, t, v) = \lambda_a(z, t) W(v),$$

(6.8.5)

where $W(v)$ is the normalized velocity distribution function, which can be taken as a Maxwellian distribution,

$$W(v) = \frac{\exp(-v^2/2 \langle v^2 \rangle)}{\sqrt{2\pi \langle v^2 \rangle}}, \quad \tfrac{1}{2}m \langle v^2 \rangle = \tfrac{1}{2}\varkappa T.$$

(6.8.6)

One then finds that (6.6.8) is replaced by

$$N(z, v) = \varrho_{aa}(z, v) - \varrho_{bb}(z, v) = \frac{W(v) N^{(0)}(z)}{1 + (R(z, v)/R_s)},$$

(6.8.7)

which represents the density of population inversion for atoms with velocity v.

By (6.8.4), the induced transition rate $R(z, v)$ in mode n has resonant maxima for speeds v such that

$$\omega_n - \omega_0 = \pm k_n v.$$

(6.8.8)

Therefore, by the saturation effects contained in (6.8.7), the inversion density as a function of v will have "holes" corresponding to atoms moving at these speeds. As shown by the Lorentzian factors in (6.8.4), the width of these holes is given by the homogeneous linewidth γ_{ab},

$$\Delta v \sim \gamma_{ab}/k_n. \tag{6.8.9}$$

One usually refers to this effect as "hole-burning" in the velocity distribution (Fig. 6.8). In particular, for a mode at resonance with the atomic transition, $\omega_n = \omega_0$, the two holes coalesce into a single deeper one at $v = 0$.

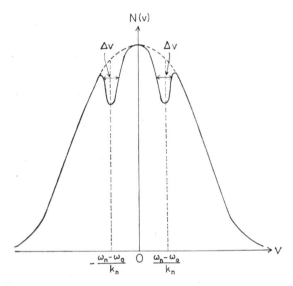

Figure 6.8 "Hole burning".

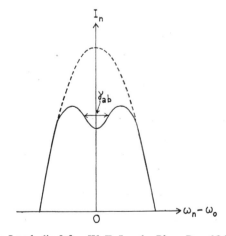

Figure 6.9 The Lamb dip [after W. E. Lamb, *Phys. Rev.* **134**, A 1429 (1964)].

From (6.8.7), by a procedure similar to that employed in the previous treatment, one can determine the coefficients $C_n(v)$ and $S_n(v)$, and C_n and S_n are then obtained by integrating over all velocities. The results are similar to those obtained in the previous treatment, except that the atomic line shape factor now contains the effect of Doppler broadening. Lamb's original paper[14] should be consulted for additional details.

Under certain conditions, the "hole" at $v = 0$ at resonance can be observed in the curve for the intensity of excitation I_n of mode n as a function of the "detuning" $\omega_n - \omega_0$ (Fig. 6.9). This effect, known as the "Lamb dip", has been observed experimentally.[15]

The effects of mode competition (§ 6.7) have also been observed,[16] and the results are in good agreement with Lamb's theory.

References

1. W. E. Lamb, *Phys. Rev.* **134**, A 1429 (1964); also, *QOE* (see Ch. 1, Ref. 2), p. 331.
2. W. E. Lamb, in *Lectures in Theoretical Physics*, vol. 2, eds. W. E. Brittin and B. W. Downs, Interscience, New York (1960), p. 435.
3. W. E. Lamb, *QECL* (see Ch. 1, Ref. 1), p. 78.
4. I. I. Rabi, *Phys. Rev.* **51**, 653 (1937).
5. N. Krylov and N. Bogoliubov, *Introduction to Nonlinear Mechanics*, Princeton Univ. Press (1943).
6. Note that $\dot{\varphi}_n$ is not necessarily negligible in (6.2.19) because it has to be compared with the small quantity $\omega_n - \Omega_n$ [cf. (6.2.9)].
7. V. F. Weisskopf and E. P. Wigner, *Z. Physik* **63**, 54 (1930); W. Heitler, *The Quantum Theory of Radiation*, 3rd ed., Oxford Univ. Press (1954), p. 182.
8. R. P. Feynman, F. L. Vernon, Jr., and R. W. Hellwarth, *J. Appl. Phys.* **28**, 49 (1957). Cf., also, R. P. Feynman, R. B. Leighton, and M. Sands, *The Feynman Lectures in Physics*, vol. III, Addison-Wesley, Reading (1964).
9. W. E. Lamb, in *Lectures in Theoretical Physics*, vol. 2, eds. W. E. Brittin and B. W. Downs, Interscience, New York (1960), p. 477.
10. W. Heitler, *The Quantum Theory of Radiation*, 3rd ed., Oxford Univ. Press (1954), p. 33.
11. Cf. W. Heitler, ibid., pp. 25, 179.
12. B. Van der Pol, *Phil. Mag.* **43**, 700 (1922); *Proc. I.R.E.* **22**, 1051 (1934).
13. Cf. V. Volterra, *Théorie Mathématique de la Lutte pour la Vie*, Gauthier-Villars, Paris (1931); N. S. Goel, S. C. Maitra and E. W. Montroll, *Rev. Mod. Phys.* **43**, 231 (1971).
14. W. E. Lamb, *Phys. Rev.* **134**, A 1429 (1964).
15. R. A. MacFarlane, W. R. Bennett and W. E. Lamb, *Appl. Phys. Letters* **2**, 189 (1963); A. Szöke and A. Javan, *Phys. Rev. Letters* **10**, 521 (1963).
16. R. L. Fork and M. A. Pollack, *Phys. Rev.* **139**, A 1408 (1965).

CHAPTER 7

Quantum Theory of the Laser

7.1 INTRODUCTION

As HAS ALREADY been mentioned in § 6.1, the main shortcoming of the semiclassical theory is that it does not include the effect of spontaneous emission. As a consequence of this, the following problems cannot be adequately treated within the framework of the semiclassical theory:

(1) INTRINSIC LINEWIDTH For single-mode oscillation, the semiclassical treatment leads to a perfectly monochromatic laser field (cf. end of § 6.7). In practice, the linewidth of the laser field is due mainly to fluctuations in the mode frequencies due, e.g., to mechanical vibrations of the mirrors. However, even if these fluctuations were completely eliminated, there would remain an *intrinsic linewidth*, due to noise sources that cannot be eliminated: spontaneous emission, thermal noise and zero-point fluctuations. Although the intrinsic linewidth is so small that it has thus far remained inaccessible to measurements, its evaluation is of fundamental interest.

(2) TRANSIENT BUILDUP FROM VACUUM According to (6.7.19), a zero-field (vacuum) solution exists, even above threshold. Although it is unstable, the presence of an initial field (no matter how small) is required to trigger the transition into the state of stable laser oscillation. In a quantized theory, such a field is automatically provided by the effect of spontaneous emission. On the other hand, below threshold, the steady state, according to the semiclassical theory, is the zero-field solution [cf. (6.5.11)], whereas one expects that it must actually correspond to a chaotic field [§ 4.1(a)], due to random spontaneous emission by different atoms (with some amplification by induced emission). This expectation is confirmed by the quantized theory, as will be seen below.

(3) PHOTON STATISTICS The photon counting probabilities for the laser field can, of course, only be obtained in the quantized theory. According to the above discussion, we expect them to correspond to a chaotic field below threshold and to approximate a coherent field above threshold.

There are several different approaches to the quantum theory of the laser; depending on which aspects of the above problems one wants to deal with, some are more suitable than others. They are based on the same type of model (cf. Fig. 6.1), and they all lead to very similar results.

We will adopt the *density operator approach*, which is similar to that employed by Wangsness and Bloch[1] for treating the interaction between nuclear spins and heat baths in the theory of magnetic resonance. We will present the version due to Scully and Lamb[2], in a greatly simplified form[3], which nevertheless suffices to bring out the main physical ideas very clearly. This will also enable us to make a parallel with Lamb's semiclassical theory. Other approaches will be briefly reviewed later (cf. § 7.7).

There is not much point in trying to include all the complications that have already been adequately dealt with by the semiclassical theory, so that we confine our attention to the simplest possible situation and to the three problems enumerated above, which specifically require the quantized treatment. Thus, besides the assumptions (i) to (iv) of § 6.1, we now make the following additional assumptions:

(v) SINGLE-MODE OSCILLATION AT RESONANCE   We consider only the case [cf. (6.5.25)]

$$\omega_n = \Omega_n = \omega_0. \tag{7.1.1}$$

(vi) NO ATOMIC MOTION   We neglect the complications due to the Doppler effect (§ 6.8).

(vii) NO SPATIAL VARIATION OF THE MODE   We substitute $\sin(k_n z)$ in (6.2.5) by its root mean square value,

$$\sin(k_n z) \to 1/\sqrt{2}. \tag{7.1.2}$$

Even with these simplifying assumptions, however, the problem is still too difficult to have been solved in full generality. We will discuss the solution only under the condition that another basic assumption be satisfied:

(viii) ADIABATIC APPROXIMATION   We assume that *the field is slowly-varying*, so that it remains practically constant during an effective atomic lifetime T, where [cf. (6.4.21)]

$$T \sim 1/\gamma_{ab} \tag{7.1.3}$$

is the time it takes for each active atom to give its contribution to the field. Typically, $T \sim 10^{-8}$ s (taking into account collision broadening). On the other hand, the characteristic time interval τ_f required for an appreciable variation in the field, in the semiclassical theory, would correspond roughly to the exponential rise time in the linear approximation, according to (6.5.11); typically, $\tau_f \gtrsim 10^{-6}$ s. Thus,

$$T \ll \tau_f, \tag{7.1.4}$$

which is the justification for the adiabatic approximation. The field is the resultant of a large number of small contributions due to different atoms, and we can neglect the change in the field during the time T of each individual contribution, so that the active atoms follow the evolution of the field adiabatically.

The basic model for the laser has already been described in § 6.1 (cf. Fig. 6.1). However, we must now be more careful in the treatment of the loss reservoirs. In the semiclassical theory, the atomic loss reservoirs were introduced phenomenologically through the replacement of real frequencies by complex frequencies [cf. (6.3.8)], even though this violated the hermiticity of the hamiltonian. In the quantized theory, such a replacement would have even more drastic effects. Thus, in the equation of motion (3.1.19) for the annihilation operator of a field mode in the Heisenberg picture, it would lead to [cf. (6.2.23)]

$$a(t) \rightarrow a \exp\left(-i\omega t - \gamma t/2\right), \tag{7.1.5}$$

and, for large times, this would violate[4] the commutation relations (2.1.8).

For a consistent description of damping in the quantum theory[5], we have to keep track of the fact that the energy lost by the damped system is distributed among the degrees of freedom of the loss reservoir. In the limit as the number of degrees of freedom tends to infinity, we get irreversible behavior. It is important to notice that we do not observe the coordinates associated with the reservoir degrees of freedom: we confine our attention only to the system of interest. We now describe the procedure that is employed to get rid of the unobserved coordinates, known as the *reduction of the density operator*.

7.2 THE REDUCED DENSITY OPERATOR

Let us assume that we have two interacting systems A and B, of which B plays the role of a reservoir: we are not interested in following the evolution of system B, but only in how the evolution of A is affected by its interaction with B.

In classical statistical mechanics, we can define a joint probability density in phase space, dependent on the coordinates and momenta of both systems, $\varrho_{AB}(p_A, q_A; p_B, q_B)$, such that the statistical average of a physical quantity $\mathcal{O}$ associated with both systems is given by [cf. (2.1.18)]

$$\overline{\mathcal{O}} = \int \varrho_{AB}\, \mathcal{O} \, dp_A \, dq_A \, dp_B \, dq_B. \tag{7.2.1}$$

In particular, for a quantity $\mathscr{O}_A$ associated only with the system of interest A (i.e., depending only on p_A, q_A), this becomes

$$\overline{\mathscr{O}}_A = \int \varrho_A(p_A, q_A)\, \mathscr{O}_A\, dp_A\, dq_A, \tag{7.2.2}$$

where

$$\varrho_A(p_A, q_A) = \int \varrho_{AB}(p_A, q_A; p_B, q_B)\, dp_B\, dq_B \tag{7.2.3}$$

is the probability density associated only with system A (regardless of the values of the B coordinates).

Similarly, in quantum statistical mechanics, let ϱ_{AB} be the density operator of the combined system, which may be represented by the density matrix

$$\varrho_{AB} = \varrho_{\alpha\beta, \alpha'\beta'}, \tag{7.2.4}$$

where α, α' are indices (or sets of indices) associated with system A and β, β' are associated with the B variables.

We are interested only in the expectation values of A-system operators $\mathscr{O}_A$, which act only on the A variables α, α'. Such an operator may be regarded as an operator on the combined system by taking its direct product with the identity operator for system B,

$$\mathscr{O}_A \otimes \mathbb{1}_B = \mathscr{O}_{\alpha, \alpha'} \delta_{\beta, \beta'}. \tag{7.2.5}$$

The expectation value of $\mathscr{O}_A$ becomes

$$\langle \mathscr{O}_A \rangle = \mathrm{Tr}\,(\varrho_{AB} \mathscr{O}_A \otimes \mathbb{1}_B) = \sum_{\alpha, \alpha'} \sum_{\beta, \beta'} \varrho_{\alpha\beta, \alpha'\beta'} \mathscr{O}_{\alpha'\alpha} \delta_{\beta'\beta}$$

$$= \sum_{\alpha, \alpha'} \mathscr{O}_{\alpha'\alpha} \sum_{\beta\beta'} \varrho_{\alpha\beta, \alpha'\beta'} \delta_{\beta'\beta} = \sum_{\alpha, \alpha'} \mathscr{O}_{\alpha', \alpha} (\varrho_A)_{\alpha\alpha'},$$

i.e.,

$$\langle \mathscr{O}_A \rangle = \mathrm{Tr}_A\,(\varrho_A \mathscr{O}_A), \tag{7.2.6}$$

where

$$(\varrho_A)_{\alpha\alpha'} = \sum_{\beta} \varrho_{\alpha\beta, \alpha'\beta} = (\mathrm{Tr}_B\, \varrho_{AB})_{\alpha\alpha'},$$

i.e.,

$$\varrho_A = \mathrm{Tr}_B(\varrho_{AB}). \tag{7.2.7}$$

The results (7.2.6), (7.2.7) are the quantum analogues of (7.2.2), (7.2.3). The operator ϱ_A, which represents the information relative to system A alone contained in ϱ_{AB}, is called the *reduced density operator*; it is the projection of ϱ_{AB} on the subspace corresponding to system A.

The systems A and B are *independent* (uncorrelated) when

$$\langle \mathscr{O}_A \mathscr{O}_B \rangle = \langle \mathscr{O}_A \rangle \langle \mathscr{O}_B \rangle \tag{7.2.8}$$

for all pairs of operators $\mathscr{O}_A$ (acting only on A) and $\mathscr{O}_B$ (acting only on B). This implies a *factorization* of ϱ_{AB},

$$\varrho_{AB} = \varrho_A \varrho_B. \tag{7.2.9}$$

Even if A and B are independent at $t = 0$, and ϱ_A at $t = 0$ represents a pure state of A, the interaction with B introduces correlations between the two systems and, at a later time t, ϱ_A in general represents a mixed state. This loss of information results from the fact that the evolution of B is not observed. The reduction of the density operator thus allows one to describe an irreversible process for system A, as well as to introduce "random noise" effects[6].

In the quantum theory of the laser, the reduction of the density operator is employed several times: (a) For the quantum treatment of damping, i.e., to describe the effect of the atom and field loss reservoirs in Fig. 6.1. For this purpose, one begins by considering the density operator for the combined system of atom (or field) and reservoir, and then one traces over the reservoir variables [cf. (7.2.7)]. (b) We are ultimately interested only in the behavior of the laser field, and not in following the evolution of the active atoms. Thus, we consider first the density operator ϱ_{FA} of the laser system (field + active atoms) in Fig. 6.1 (with loss reservoir variables already traced over), and from it we derive the reduced density operator ϱ_F for the laser field alone (for brevity, we shall later omit the index F) by tracing over the atomic variables,

$$\varrho_F = \mathrm{Tr}_A \left(\varrho_{FA} \right). \tag{7.2.10}$$

The effects of noise (including that due to spontaneous emission) are introduced by means of this reduction process.

7.3 LASER MODEL AND HAMILTONIAN

(a) The laser model

With reference to the laser model of Fig. 6.1, we consider a gas laser in single-mode oscillation at resonance. We know from the semiclassical theory that, in order to describe steady-state oscillation, we need a *nonlinear active medium* to supply energy to the field. The active medium will again be idealized as a collection of 2-level atoms which are pumped to the upper level a at random times t_{oi}, with pumping rate λ_a per unit volume [we take $\lambda_b = 0$ for simplicity; cf. (6.4.12)].

In the treatment of Scully and Lamb, the atom loss reservoirs are introduced by assuming that a and b decay to lower levels c and d with the emission of (nonlaser) radiation, and decay constants γ_a and γ_b [cf. (6.3.8)]. Scully and Lamb give a detailed treatment of this four-level system interacting with the radiation field in the Weisskopf–Wigner approximation, from which they derive the damping terms in the equations of motion for the density operator, by tracing over reservoir variables.

Following Scully[7], we simplify the treatment by assuming that a given atom, pumped to level a at a time t_0, is removed from the system at time $t_0 + T$, after interacting with the field during an effective atomic lifetime T [cf. (7.1.3)], in which it gives its total contribution to the field.

Furthermore, we assume T to be small enough that the contribution

$$\delta \varrho_F = \varrho_F(t_0 + T) - \varrho_F(t_0) \qquad (7.3.1)$$

given by *one* atom to the reduced density operator for the field may be computed by time-dependent perturbation theory. This is not really consistent. It would correspond to substituting the Weisskopf–Wigner exponential decay approximation by a perturbation series,

$$e^{-\gamma t} = 1 - \gamma t + \frac{\gamma^2}{2!} t^2 - \cdots, \qquad (7.3.2)$$

and we want to do this for times $t = T \gtrsim 1/\gamma$, for which the series is not rapidly convergent. However, the perturbation approach is considerably simpler, and it is justified by comparing the results with those obtained by Scully and Lamb in the Weisskopf–Wigner approximation: it turns out that the *form* of the equations of motion is the same, the only difference lying in the definition of the coefficients that appear in the equations.

If we carry the perturbation series up to second order, this turns out to correspond to the linear approximation; to describe a nonlinear active medium, we have to retain the next nonvanishing contribution; as will be seen later, this requires us to carry the perturbation series up to fourth order.

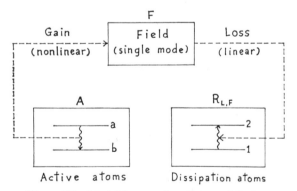

Figure 7.1   Model for quantum theory of the laser

It is again *assumed*, as in the semiclassical treatment, that the precise details of the loss mechanisms do not play an important role in the theory (some justification for this is provided by experimental evidence, as well as by the fact that different theoretical models lead to similar results; however, a more detailed investigation of this point might be of interest). To represent the field loss reservoir $R_{L,F}$ (Fig. 7.1), Scully and Lamb take another set of (nonresonant) two-level atoms that are injected at random times in the *lower* level 1, so as to absorb energy from the laser field, making transitions to the upper level 2. The advantage of this model is that it can be treated by the same method applied to the active atoms, except that the initial conditions are different for the dissipation atoms. The linear approximation is sufficient in this case, so that we can stop at second order in the simplified perturbation treament.

(b) Field quantization

We expand the field in terms of axial modes [cf. (5.2.4), (6.2.5)]. In the Schrödinger picture, because of assumption (v) (single mode present), the quantized electric field takes the form

$$E(z) = \mathscr{E}(a + a^+) \sin(kz), \tag{7.3.3}$$

where E is the x-component of the field (assumed linearly polarized), a and a^+ are the annihilation and creation operators, and the mode index has been dropped for simplicity. The normalization factor $\mathscr{E}$ is taken to be

$$\mathscr{E} = (\hbar\Omega/\varepsilon_0 V)^{1/2}, \tag{7.3.4}$$

where $\Omega = ck$ is the frequency and V is the volume of the laser.

Let ϱ [previously denoted by ϱ_F in (7.2.10)] be the reduced density operator for the laser field. In the Fock representation,

$$\varrho = \sum_{n,n'} \varrho_{n,n'} |n\rangle \langle n'|, \tag{7.3.5}$$

and (cf. § 4.2)

$$p(n) = \varrho_{n,n} \tag{7.3.6}$$

defines the probability of finding n photons in the field (photon statistics).

The expectation value of E is

$$\langle E \rangle = \mathscr{E} \sin(kz) \, \mathrm{Tr} \, [\varrho(a + a^+)]$$

$$= \mathscr{E} \sin(kz) \sum_{n,n'} \varrho_{n,n'} \langle n' | (a + a^+) | n \rangle$$

$$= \mathscr{E} \sin(kz) \sum_{n} (\sqrt{n + 1} \, \varrho_{n,n+1} + c.c.), \tag{7.3.7}$$

where c.c. denotes the complex conjugate, and we have employed (3.1.14), (3.1.15).

Note that, while the photon statistics (7.3.6) depends only on the diagonal elements of ϱ in the Fock representation, the expectation value of the electric field (7.3.7) depends only on the elements on the first parallel above and below the main diagonal. In particular, if ϱ is diagonal in the Fock basis, $\langle E \rangle = 0$. Examples in which this happens are: (i) *Fock state*, $\varrho = |n\rangle \langle n|$; (ii) *Thermal field*; in this case, by (4.1.5),

$$\varrho_{n,n'} = (1 - \zeta) \zeta^n \delta_{n,n'}, \quad \zeta = e^{-\beta \hbar \Omega}. \tag{7.3.8}$$

On the other hand, for the case (iii) *Pure coherent state*, we have, by (3.1.22), (3.1.23),

$$\varrho = |v\rangle \langle v|, \quad \varrho_{n,n'} = \exp(-|v|^2) \frac{v^n (v^*)^{n'}}{\sqrt{n! n'!}}, \tag{7.3.9}$$

and (7.3.7) yields

$$\langle E \rangle = \mathscr{E} \sin(kz)(v + v^*) = 2\mathscr{E} \sin(kz) |v| \cos \varphi. \tag{7.3.10}$$

However, if we take (iv) *Uniform phase average of coherent state*, as in (4.1.21), we get

$$\varrho = \frac{1}{2\pi} \int_0^{2\pi} ||v| e^{i\varphi}\rangle \langle |v| e^{i\varphi}| \, d\varphi, \quad \varrho_{n,n'} = \exp(-|v|^2) \frac{|v|^{2n}}{n!} \delta_{n,n'}, \tag{7.3.11}$$

so that again $\langle E \rangle = 0$ in this case.

(c) Hamiltonian

Neglecting zero-point energy, the field hamiltonian with a single mode present is

$$\mathscr{H}_F = \hbar H_F = \hbar \omega a^+ a, \tag{7.3.12}$$

where we have adopted the notation ω for the single frequency $\omega = \Omega = \omega_0$ [cf. (7.1.1)].

Similarly, the free hamiltonian for the active atoms [cf. § 6.3(a)] is given by

$$\mathscr{H}_A = \hbar H_A = \hbar \begin{pmatrix} \omega_a & 0 \\ 0 & \omega_b \end{pmatrix}. \tag{7.3.13}$$

In terms of the raising and lowering (spin-flip) operators

$$\sigma_{\pm} = \tfrac{1}{2}(\sigma_1 \pm i\sigma_2), \quad \sigma_+ = \begin{pmatrix} 0 & 1 \\ 0 & 0 \end{pmatrix}, \quad \sigma_- = \begin{pmatrix} 0 & 0 \\ 1 & 0 \end{pmatrix}, \tag{7.3.14}$$

we may rewrite (7.3.13) as

$$H_A = \omega_a \sigma_+ \sigma_- + \omega_b \sigma_- \sigma_+. \tag{7.3.15}$$

The interaction hamiltonian is still given by (6.3.21),

$$\mathscr{V} = \hbar V = \hbar \begin{pmatrix} 0 & V_{ab} \\ V_{ab} & 0 \end{pmatrix} = \hbar V_{ab}(\sigma_+ + \sigma_-), \tag{7.3.16}$$

but V_{ab} [cf. (6.3.16)] now contains the quantized field,

$$V_{ab} = -\frac{d}{\hbar} E = -\frac{e}{\hbar} \langle x_{ab} \rangle E, \tag{7.3.17}$$

where E is given by (7.3.3). We now make use of assumption (vii), by making the substitution (7.1.2),

$$E \rightarrow \frac{\mathscr{E}}{\sqrt{2}} (a + a^+), \tag{7.3.18}$$

so that (7.3.17) becomes

where

$$V_{ab} = g(a + a^+), \tag{7.3.19}$$

$$g = -e\mathscr{E} \langle x_{ab} \rangle / (\sqrt{2} \, \hbar) \tag{7.3.20}$$

plays the role of a *coupling constant*, which is linear in the field amplitude $\mathscr{E}$. The interaction hamiltonian (7.3.16) becomes

$$V = g(a + a^+)(\sigma_+ + \sigma_-). \tag{7.3.21}$$

Finally, the total hamiltonian of the laser system is given by

$$\mathscr{H} = \hbar H, \tag{7.3.22}$$

where

$$\begin{aligned}
H &= H_F + H_A + V = H_0 + V \\
&= \omega a^+ a + \omega_a \sigma_+ \sigma_- + \omega_b \sigma_- \sigma_+ + g(a + a^+)(\sigma_+ + \sigma_-).
\end{aligned} \tag{7.3.23}$$

7.4 EQUATIONS OF MOTION

(a) The combined system

The equation of motion for the density operator of the combined system (field interacting with a two-level atom) is

$$\dot{\varrho}_{FA} = -i[(H_0 + V), \varrho_{FA}]. \tag{7.4.1}$$

Let us go over to the interaction picture [cf. (2.2.8)], with

$$\mathcal{O}^{(I)} = e^{iH_0 t} \mathcal{O}^{(S)} e^{-iH_0 t} \tag{7.4.2}$$

relating operators in the interaction (I) and Schrödinger (S) pictures. The equation of motion (7.4.1) becomes

$$\dot{\varrho}_{FA}^{(I)} = -i[V^{(I)}, \varrho_{FA}^{(I)}]. \tag{7.4.3}$$

By (7.3.23) and (7.4.2), we have

$$V^{(I)} = e^{iH_0 t} V^{(S)} e^{-iH_0 t} = g e^{i\omega a^+ a t} (a + a^+) e^{-i\omega a^+ a t}$$

$$\times \exp\left[i(\omega_a \sigma_+ \sigma_- + \omega_b \sigma_- \sigma_+) t \right] (\sigma_+ + \sigma_-) \exp\left[-i(\omega_a \sigma_+ \sigma_- + \omega_b \sigma_- \sigma_+) t \right]. \tag{7.4.4}$$

By using the relation (3.2.5), together with

$$[a^+a, a] = -a,$$

we get

$$e^{i\omega a^+ at}ae^{-i\omega a^+ at} = \left[1 - i\omega t + \frac{(-i\omega t)^2}{2!} + \cdots\right]a = e^{-i\omega t}a. \quad (7.4.5)$$

Similarly,

$$\exp\left[i(\omega_a\sigma_+\sigma_- + \omega_b\sigma_-\sigma_+)t\right](\sigma_+ + \sigma_-)\exp\left[-i(\omega_a\sigma_+\sigma_- + \omega_b\sigma_-\sigma_+)t\right]$$

$$= \begin{pmatrix} e^{i\omega_a t} & 0 \\ 0 & e^{i\omega_b t} \end{pmatrix}\begin{pmatrix} 0 & 1 \\ 1 & 0 \end{pmatrix}\begin{pmatrix} e^{-i\omega_a t} & 0 \\ 0 & e^{-i\omega_b t} \end{pmatrix} = \begin{pmatrix} 0 & e^{i\omega_0 t} \\ e^{-i\omega_0 t} & 0 \end{pmatrix}$$

$$= e^{i\omega_0 t}\sigma_+ + e^{-i\omega_0 t}\sigma_-, \quad (7.4.6)$$

where ω_0 is given by (6.3.3).

Substituting (7.4.5) and (7.4.6) in (7.4.4), we get

$$V^{(I)} = g(ae^{-i\omega t} + a^+e^{i\omega t})(\sigma_+e^{i\omega_0 t} + \sigma_-e^{-i\omega_0 t})$$

$$= g[\sigma_+ae^{-i(\omega-\omega_0)t} + \sigma_-a^+e^{i(\omega-\omega_0)t} + \text{antiresonant terms}]. \quad (7.4.7)$$

We now make the rotating-wave approximation [cf. (6.1.1)] by neglecting the antiresonant terms. Taking into account the assumption (v) of exact resonance [cf. (7.1.1)], (7.4.7) becomes

$$V^{(I)} = g(\sigma_+a + \sigma_-a^+) = g\begin{pmatrix} 0 & a \\ a^+ & 0 \end{pmatrix}. \quad (7.4.8)$$

We can also say that the elementary process described by the first term of (7.4.8) is an atomic transition from b to a accompanied by the absorption of a photon, whereas the second term describes the reverse process, and we neglect terms in which the atom makes a downward (upward) transition, accompanied by absorption (emission) of a photon.

From now on we work in the interaction picture, omitting the index (I) for brevity. The equation of motion (7.4.3), together with an initial condition, is equivalent to the integral equation

$$\varrho_{FA}(t) = \varrho_{FA}(t_0) - i\int_{t_0}^{t} [V(t'), \varrho_{FA}(t')]\,dt'. \quad (7.4.9)$$

(b) Perturbation solution

We are interested in computing the contribution of a single atom to the reduced density operator for the field, i.e., by (7.3.1) and (7.2.10),

$$\delta\varrho_F = \text{Tr}_A\left[\varrho_{FA}(t_0 + T)\right] - \text{Tr}_A[\varrho_{FA}(t_0)]. \quad (7.4.10)$$

At the time of injection t_0, the atom and the field are uncorrelated, so that

$$\varrho_{FA}(t_0) = \varrho_0 \varrho_A(t_0), \tag{7.4.11}$$

where

$$\varrho_0 = \varrho_F(t_0) \tag{7.4.12}$$

and [cf. (6.3.25)]

$$\varrho_A(t_0) = |a\rangle \langle a| = \begin{pmatrix} 1 & 0 \\ 0 & 0 \end{pmatrix}, \tag{7.4.13}$$

so that

$$\varrho_{FA}(t_0) = \begin{pmatrix} \varrho_0 & 0 \\ 0 & 0 \end{pmatrix}. \tag{7.4.14}$$

As explained in § 7.3(a), we solve (7.4.9) by time-dependent perturbation theory,

$$\varrho_{FA}(t) = \varrho_{FA}^{(0)}(t) + \varrho_{FA}^{(1)}(t) + \varrho_{FA}^{(2)}(t) + \cdots, \tag{7.4.15}$$

where

$$\varrho_{FA}^{(0)}(t) = \varrho_{FA}(t_0), \tag{7.4.16}$$

$$\varrho_{FA}^{(n+1)}(t) = -i \int_{t_0}^{t} [V(t'), \varrho_{FA}^{(n)}(t')] \, dt' \quad (n = 0, 1, 2, \ldots). \tag{7.4.17}$$

Substituting (7.4.14) and (7.4.8) in (7.4.16) and (7.4.17), we find, successively,

$$\varrho_{FA}^{(1)}(t) = -ig \int_{t_0}^{t} dt' \left[\begin{pmatrix} 0 & a \\ a^+ & 0 \end{pmatrix}, \begin{pmatrix} \varrho_0 & 0 \\ 0 & 0 \end{pmatrix} \right]$$

$$= -ig(t - t_0) \begin{pmatrix} 0 & -\varrho_0 a \\ a^+ \varrho_0 & 0 \end{pmatrix}, \tag{7.4.18}$$

$$\varrho_{FA}^{(2)}(t) = \frac{(-ig)^2}{2!} (t - t_0)^2 \begin{pmatrix} aa^+ \varrho_0 + \varrho_0 aa^+ & 0 \\ 0 & -2a^+ \varrho_0 a \end{pmatrix}, \tag{7.4.19}$$

$$\varrho_{FA}^{(3)}(t) = \frac{(-ig)^3}{3!} (t - t_0)^3 \begin{pmatrix} 0 & -\varrho_0 aa^+ a - 3aa^+ \varrho_0 a \\ a^+ aa^+ \varrho_0 + 3a^+ \varrho_0 aa^+ & 0 \end{pmatrix}, \tag{7.4.20}$$

$$\varrho_{FA}^{(4)}(t) = \frac{(-ig)^4}{4!} (t - t_0)^4$$

$$\times \begin{pmatrix} aa^+ aa^+ \varrho_0 + 6aa^+ \varrho_0 aa^+ + \varrho_0 aa^+ aa^+ & 0 \\ 0 & -4a^+ \varrho_0 aa^+ a - 4a^+ aa^+ \varrho_0 a \end{pmatrix}. \tag{7.4.21}$$

10*

Since the trace over the atomic variables corresponds to the sum of the diagonal elements, we see that only even-order terms contribute to (7.4.10). Retaining contributions up to fourth order in (7.4.10), we find, for the field gain due to a single atom,

$$(\delta\varrho_F)_{\text{gain}} = -\frac{g^2}{2!} T^2(aa^+\varrho_0 + \varrho_0 aa^+ - 2a^+\varrho_0 a)$$

$$+\frac{g^4}{4!} T^4(aa^+aa^+\varrho_0 + 6aa^+\varrho_0 aa^+ + \varrho_0 aa^+aa^+ - 4a^+\varrho_0 aa^+a - 4a^+aa^+\varrho_0 a).$$

$$(7.4.22)$$

As explained in § 7.3(a), the effect due to the interaction with a single dissipation atom in Fig. 7.1 can be derived in exactly the same way, except that the atom is now injected in the lower level, so that (7.4.13) is replaced by

$$\varrho_A(t_0) = |b\rangle \langle b| = \begin{pmatrix} 0 & 0 \\ 0 & 1 \end{pmatrix}. \tag{7.4.23}$$

Furthermore, since the dissipation mechanism is linear, it suffices to keep the lowest nonvanishing order of perturbation theory, i.e., second order. We then find, in the place of (7.4.22),

$$(\delta\varrho_F)_{\text{loss}} = -\frac{g'^2}{2!} T^2(a^+a\varrho_0 + \varrho_0 a^+a - 2a\varrho_0 a^+). \tag{7.4.24}$$

(c) Reduced density operator for the field

We now use the basic adiabatic assumption (viii) of § 7.1. It follows from (7.1.4), in the first place, that

$$\varrho_F(t) \approx \varrho_F(t_0) = \varrho_0 \quad (t_0 \leqq t \leqq t_0 + T). \tag{7.4.25}$$

On the other hand, we can consider a time interval Δt such that

$$T \lesssim \Delta t \ll \tau_f, \tag{7.4.26}$$

during which

$$r_a\Delta t = \lambda_a V\Delta t \tag{7.4.27}$$

atoms are pumped to level a (V = volume of the laser). The total contribution to the reduced density operator for the field due to these atoms is

$$(\Delta\varrho_F)_{\text{gain}} = r_a\Delta t(\delta\varrho_F)_{\text{gain}}, \tag{7.4.28}$$

where $(\delta\varrho_F)_{\text{gain}}$ is given by (7.4.22).

We may now identify

$$\left(\frac{d\varrho_F}{dt}\right)_{\text{gain}} = \left(\frac{\Delta\varrho_F}{\Delta t}\right)_{\text{gain}} = r_a(\delta\varrho_F)_{\text{gain}} = -\frac{A}{2}(aa^+\varrho_F + \varrho_F aa^+ - 2a^+\varrho_F a)$$

$$+\frac{B}{8}(aa^+aa^+\varrho_F + 6aa^+\varrho_F aa^+ + \varrho_F aa^+aa^+ - 4a^+\varrho_F aa^+a - 4a^+aa^+\varrho_F a),$$

$$(7.4.29)$$

where

$$A = r_a g^2 T^2, \tag{7.4.30}$$

$$B = 8r_a g^4 T^4/4!. \tag{7.4.31}$$

Similarly, from (7.4.24), we get

$$\left(\frac{d\varrho_F}{dt}\right)_{\text{loss}} = -\frac{C}{2}(a^+a\varrho_F + \varrho_F a^+a - 2a\varrho_F a^+). \tag{7.4.32}$$

We identify the parameter C with the rate of energy loss per unit time [cf. (5.2.6)]

$$C = \omega/Q; \tag{7.4.33}$$

this identification is justified by the results derived below.

Putting together (7.4.29) and (7.4.32), and dropping the index F for brevity, we finally obtain the equation of motion for the reduced density operator of the laser field:

$$\frac{d\varrho}{dt} = \left(\frac{d\varrho}{dt}\right)_{\text{gain}} + \left(\frac{d\varrho}{dt}\right)_{\text{loss}} = -\frac{A}{2}(aa^+\varrho + \varrho aa^+ - 2a^+\varrho a)$$

$$+\frac{B}{8}(aa^+aa^+\varrho + 6aa^+\varrho aa^+ + \varrho aa^+aa^+ - 4a^+\varrho aa^+a - 4a^+aa^+\varrho a)$$

$$-\frac{C}{2}(a^+a\varrho + \varrho a^+a - 2a\varrho a^+). \tag{7.4.34}$$

From this operator equation, we can find the equation for the elements of the density matrix in any representation. To discuss the photon statistics, according to (7.3.6), it is convenient to adopt the Fock representation. The equation of motion for $\varrho_{n,n'} = \langle n|\varrho|n'\rangle$ can readily be derived from (7.4.34), with the help of (3.1.14) and (3.1.15). The result is

$$\dot{\varrho}_{n,n'} = -[(n+1)R_{n,n'} + (n'+1)R_{n',n}]\varrho_{n,n'}$$

$$+ (R_{n-1,n'-1} + R_{n'-1,n-1})\sqrt{nn'}\,\varrho_{n-1,n'-1}$$

$$-\frac{C}{2}(n+n')\varrho_{n,n'} + C\sqrt{(n+1)(n'+1)}\,\varrho_{n+1,n'+1}, \tag{7.4.35}$$

where

$$R_{n,n'} = \frac{A}{2} - \frac{B}{8} [3(n' + 1) + (n + 1)].$$ (7.4.36)

The more exact calculation of Scully and Lamb, based on the Weisskopf-Wigner approximation for a four-level system instead of perturbation theory, leads to an equation of motion of precisely the same form[8], but with $R_{n,n'}$ given by

$$R_{n,n'}$$
$$= \frac{r_a g^2 \gamma_b(\gamma_{ab} + i\Delta) + g^2(n' - n)}{\gamma_a \gamma_b(\gamma_{ab}^2 + \Delta^2) + 2\gamma_{ab}^2 g^2(n + 1 + n' + 1) + g^2(n' - n)[g^2(n' - n) + i\Delta(\gamma_a - \gamma_b)]},$$ (7.4.37)

where $\gamma_{ab} = \frac{1}{2}(\gamma_a + \gamma_b)$, as in (6.4.16), and the calculation was done for the more general, nonresonant case, with $\Delta = \omega - \omega_0$.

If we set

$$\Delta = 0, \quad \gamma_a = \gamma_b = \gamma_{ab} = \gamma$$ (7.4.38)

in (7.4.37), and then expand the result in a power series in $\frac{1}{\gamma}$, we get

$$R_{n,n'} = r_a(g/\gamma)^2 - r_a(g/\gamma)^4 [3(n' + 1) + (n + 1)] + \cdots.$$ (7.4.39)

Comparing this with (7.4.36) [cf. (7.4.30), (7.4.31)], we see that, if we identify

$$T = \gamma^{-1},$$ (7.4.40)

as would be suggested by (7.3.2), there is a correspondence between the two results, provided that we identify γ^{-m} with $T^m/m!$ in m^{th} order of perturbation theory ($m = 2,4$).

The equation of motion (7.4.35) has the characteristic form of a Pauli master equation[9]. It is important to note that the elements $\varrho_{n,n'}$ of the density matrix that are coupled by this equation all have the same degree of "off-diagonality"

$$p = n - n',$$ (7.4.41)

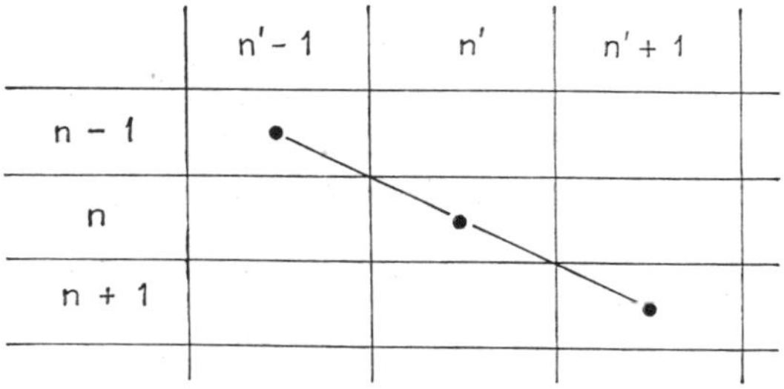

Figure 7.2 Structure of the equation of motion for the elements of the density matrix. Only elements lying along parallels to the main diagonal are coupled

so that they lie along a parallel to the main diagonal (Fig. 7.2). It follows that we can discuss the solutions separately for each value of p; in particular, the diagonal ($p = 0$) elements evolve independently of one another. We begin with the diagonal equations, which determine the photon statistics [cf. (7.3.6)].

7.5 PHOTON STATISTICS

Taking $n = n'$ in (7.4.35), with $R_{n,n'}$ given by the more exact result (7.4.37) for $\Delta = 0$, we find

$$\dot{\varrho}_{n,n} = -\frac{(n+1)A}{1+(n+1)(B/A)}\varrho_{n,n} + \frac{nA}{1+n(B/A)}\varrho_{n-1,n-1}$$

$$-nC\varrho_{n,n} + (n+1)C\varrho_{n+1,n+1}, \tag{7.5.1}$$

where

$$A = 2r_a g^2/(\gamma_a\gamma_{ab}), \tag{7.5.2}$$

$$B = 8r_a g^4/(\gamma_a^2\gamma_b\gamma_{ab}), \tag{7.5.3}$$

$$C = \omega/Q. \tag{7.5.4}$$

If we had taken $R_{n,n'}$ given by the perturbative result (7.4.36), the first two terms of (7.5.1) would have been replaced by

$$[-(n+1)A + (n+1)^2 B]\varrho_{n,n} + (nA - n^2 B)\varrho_{n-1,n-1}, \tag{7.5.5}$$

with A and B given by (7.4.30) and (7.4.31), respectively. This corresponds to approximating (7.5.1) by the expansion (note that $B/A = O(g^2)$)

$$\left(1 + n\frac{B}{A}\right)^{-1} \approx 1 - n\frac{B}{A}, \tag{7.5.6}$$

together with the substitutions already discussed following (7.4.40).

LINEAR APPROXIMATION   This corresponds to setting

$$B = 0. \tag{7.5.7}$$

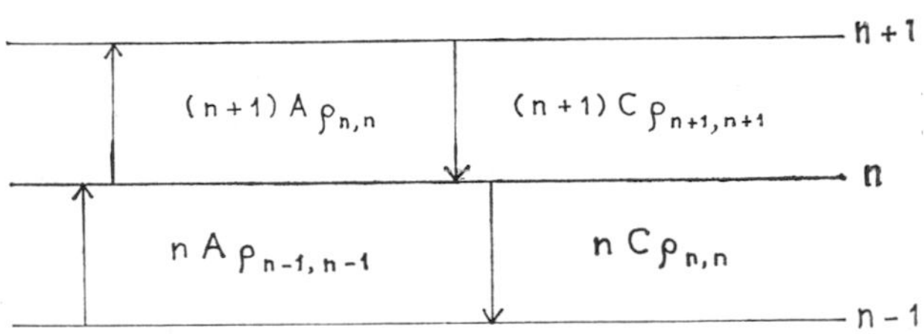

Figure 7.3 Interpretation of master equation in linear approximation

The physical interpretation of (7.5.1) in this approximation is illustrated in Fig. 7.3. The probability $p(n) = \varrho_{n,n}$ of finding n photons in the field increases by the emission of a photon by an active atom, from the state with $n - 1$ photons (term $nA\varrho_{n-1,n-1}$) and decreases by the absorption of a photon by a dissipation atom (term $-nC\varrho_{n,n}$); similarly for the level $n + 1$. The emission (absorption) term has the characteristic proportionality to the number of photons in the final (initial) state.

In this section we are interested in the *steady-state solution*, which is obtained by setting

$$\dot{\varrho}_{n,n} = 0. \tag{7.5.8}$$

This yields two identical systems of finite-difference equations, of the form

$$nA\varrho_{n-1,n-1} - nC\varrho_{n,n} = 0, \tag{7.5.9}$$

corresponding to *detailed balance* between each pair of neighboring levels (cf. Fig. 7.3). It follows that

$$\varrho_{n,n} = (A/C)\, \varrho_{n-1,n-1},$$

so that, for $A < C$, we find the solution

$$\varrho_{n,n} = \frac{(A/C)^n}{\sum_n (A/C)^n} = \left(1 - \frac{A}{C}\right)\left(\frac{A}{C}\right)^n \quad (A < C), \tag{7.5.10}$$

already normalized to yield $\mathrm{Tr}\,\varrho = 1$.

For $A \geq C$, there is no normalizable solution in the linear approximation, so that we interpret the condition

$$A = C \tag{7.5.11}$$

as defining the *treshold of oscillation*. By (7.3.20) and (7.3.4),

$$2g^2 = \frac{\omega d^2}{\varepsilon_0 \hbar V}, \tag{7.5.12}$$

so that (7.4.27), (7.5.2), (7.5.4) and (7.5.11) yield

$$\left(\frac{r_a}{V\gamma_a}\right)_c = \left(\frac{\lambda_a}{\gamma_a}\right)_c = \bar{N}_c = \frac{\varepsilon_0 \hbar \gamma_{ab}}{d^2 Q}, \tag{7.5.13}$$

where we have employed (6.6.9) (with $\lambda_b = 0$) to define the critical inversion density $\bar{N}_c$. This coincides with the result (6.5.13) obtained in the semiclassical theory.

We now see also that (7.5.10) must represent the photon statistics below threshold. Comparing this result with (7.3.8), we conclude that *below threshold, the photon statistics is the same as that for thermal radiation,* with an effective temperature T defined by

$$A/C = \zeta = \exp\left(-\hbar\omega/\varkappa T\right). \tag{7.5.14}$$

This result agrees with our expectation and corrects one of the defects of the semiclassical treatment (cf. § 7.1). It corresponds to curve (*I*) in Fig. 4.1.

NONLINEAR APPROXIMATION In order to find the steady-state solution above threshold, we must retain nonlinear terms. Let us consider first the perturbative result. According to (7.5.5), the only effect of having $B \neq 0$ as compared with the linear approximation is the substitution

$$A \to A - Bn. \tag{7.5.15}$$

The physical interpretation of the new terms is illustrated in Fig. 7.4: they correspond to a double interaction of the active atom with the field, in which it emits a photon and then reabsorbs it, thus decreasing the gain from A to $A - Bn$. Note that the effect is proportional to the number of photons present, as is characteristic of induced transitions. This is the elementary process responsible for the saturation effect.

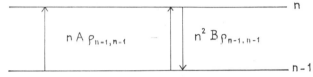

Figure 7.4 Interpretation of saturation effect in terms of probability flow

The solution (7.5.10) is now replaced by

$$\varrho_{n,n} = \mathcal{N} \prod_{j=0}^{n} \left(\frac{A - Bj}{C}\right) \quad (A > C), \tag{7.5.16}$$

where $\mathcal{N}$ is a normalization factor. Thus, $\varrho_{n,n}$ is a product of $n + 1$ factors, each of them of the form $(A - Bj)/C$. Each factor is $>1\,(<1)$ for $j < \bar{n}\,(j > \bar{n})$, where

$$\bar{n} = \frac{A - C}{B}. \tag{7.5.17}$$

It follows that $p(n) = \varrho_{n,n}$ increases monotonically with n up to $\bar{n}$ and thereafter it decreases monotonically with n, leading to a distribution peaked at $\bar{n}$, like curve (*II*) in Fig. 4.1.

The average number of photons in the field, which is a measure of the intensity, is close to $\bar{n}$:

$$\langle n \rangle \approx \bar{n} = \frac{A - C}{B} = \frac{\text{gain} - \text{loss}}{\text{saturation parameter}}. \tag{7.5.18}$$

This result should correspond to (6.7.11) of the semiclassical theory, and it provides the justification for the physical interpretation of the parameters A, B and C. In fact, $\langle n \rangle \, \hbar\omega$ must correspond to the average energy $\frac{1}{4} \, \varepsilon_0 E^2 V$ contained in the laser "cavity"; this defines the corresponding value of E^2. It is also readily seen that $A - C = 2\alpha$, where α is defined by (6.7.6), and B is proportional to β, as defined by (6.7.7).

Although (7.5.16) may take on negative values for $n > A/B$, this is a consequence only of the perturbation expansion (7.5.6) (which would then be unjustified), and it does not happen for the more exact solution, based on (7.5.1) instead of (7.5.5). In this case we find, instead of (7.5.16),

$$\varrho_{n,n} = \mathscr{N} \prod_{j=0}^{n} \frac{A/C}{[1 + (B/A)j]} , \qquad (7.5.19)$$

which reduces to (7.5.10) for $B = 0$ and to (7.5.16) for $B/A \ll 1$ and not too large n.

The more exact solution (7.5.19) shows the same monotonic, peaked behavior as (7.5.16). The peak is located at

$$\bar{n} = \frac{A}{C} \left(\frac{A - C}{B} \right) \approx \langle n \rangle , \qquad (7.5.20)$$

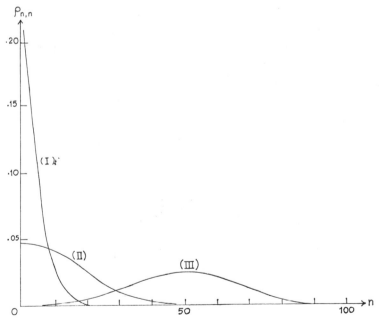

Figure 7.5 Photon statistics according to (7.5.19). (*I*) 20% below threshold. (*II*) At threshold. (*III*) 20% above threshold. The parameter B was chosen so as to yield $\langle n \rangle = 50$ [after M. O. Scully and W. E. Lamb, *Phys. Rev.* **159**, 208 (1967)]

which agrees with (7.5.18) in the region (not too far above threshold) in which the former result can be applied. The behavior of (7.5.19) in three typical operating regions of the laser is shown in Fig. 7.5. Curves (*I*) and (*III*) should be compared with those of Fig. 4.1.

In Fig. 7.6 the photon statistics given by (7.5.19) is compared with that for ideal laser light [cf. (4.2.14)], i.e., a coherent state with the same value of $\langle n \rangle$. We see the laser distribution is broader.

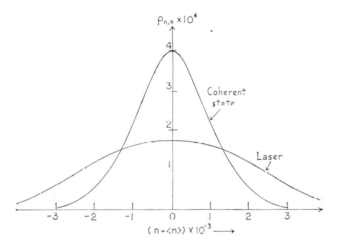

Figure 7.6 Comparison between the photon statistics for a laser 20% above threshold and for a coherent state. In both cases $\langle n \rangle = 10^6$ [after M. O. Scully and W. E. Lamb, *Phys. Rev.* **159**, 208 (1967)]

The result (7.5.19) can be rewritten in terms of the Gamma function,

$$\varrho_{n,n} = \mathcal{N} \left(\frac{A^2}{BC} \right)^n \prod_{j=0}^{n} \frac{1}{[(A/B) + j]}$$

$$= \mathcal{N} \left(\frac{A^2}{BC} \right)^n \frac{\Gamma[(A/B) + n + 1]}{\Gamma(A/B)}. \tag{7.5.21}$$

The variance of this distribution is found to be given by

$$\langle (\Delta n)^2 \rangle = \langle n^2 \rangle - \langle n \rangle^2 \approx \frac{A}{A - C} \langle n \rangle. \tag{7.5.22}$$

In particular, for a laser far above threshold ($A \gg C$), (7.5.20) becomes

$$\langle n \rangle \approx A^2/(BC), \tag{7.5.23}$$

and, in the domain $|n - \langle n \rangle| = O(\langle n \rangle^{\frac{1}{2}})$, where most of the distribution is concentrated, we have $n \gg A/B$, so that (7.5.21) can be approximated by

$$\varrho_{n,n} \approx \mathcal{N}' \frac{\langle n \rangle^n}{n!} = \exp\left(-\langle n \rangle\right) \frac{\langle n \rangle^n}{n!}, \qquad (7.5.24)$$

which is the Poisson distribution (4.2.14). Thus, the photon statistics far above threshold approaches that for a coherent state (classical limit).

The photoelectron counting statistics corresponding to a laser beam incident on a photodetector (cf. § 4.2) can be worked out[10] from (7.5.19). The experimental results[11] are in very good agreement with the theory.

7.6   TRANSIENT BEHAVIOR AND LINEWIDTH

(a) Transient buildup from vacuum

So far we have discussed only item (3) of § 7.1. In order to discuss items (1) and (2), we must look for time-dependent solutions of (7.4.35).

Beginning with the diagonal equations (7.5.1), one can look for exponentially decaying solutions of the form

$$\varrho_{n,n}(t) = \varphi(n) \, e^{-\mu t}. \qquad (7.6.1)$$

Substituting this "Ansatz" in (7.5.1), one finds a homogeneous infinite system of linear equations in infinitely many unknowns of the form

$$a_{n,n-1}\varphi(n-1) + (a_{n,n} + \mu)\,\varphi(n) + a_{n,n+1}\varphi(n+1) = 0, \qquad n = 1, 2, 3, \ldots \qquad (7.6.2)$$

The condition for a nontrivial solution is the vanishing of the determinant. The roots of the determinant form an infinite system of eigenvalues μ_j $(j = 0, 1, 2, \ldots)$. It may be shown[11a] that $\mu_j \geq 0$, and the lowest eigenvalue is

$$\mu_0 = 0, \qquad (7.6.3)$$

corresponding to the stationary solution.

The eigenfunctions $\varphi_j(n)$ corresponding to higher eigenvalues μ_j represent transient solutions, and one expects that a solution of the general form

$$\varrho_{n,n}(t) = \sum_{j=0}^{\infty} \varphi_j(n) \exp\left(-\mu_j t\right) \qquad (7.6.4)$$

allows one to describe, e.g., the transient buildup of the laser field, starting with the vacuum.

Scully, Lamb and Sargent[12] computed the solution in this case by numerical integration, taking $\langle n \rangle_{\text{steady}} = 50$ photons, $\varrho_{n,n}(t = 0) = \delta_{0,n}$

(vacuum), $\omega/Q = 0.9 \times 10^6$ s^{-1} and $A = 10^6$ s^{-1}, i.e., a laser 10% above threshold. The evolution of the photon statistics as a function of time was presented as a computer-generated movie. The form of the distribution for some values of t is shown in Fig. 7.7. The steady state is attained for $t \gtrsim 60\mu$s. The curves should be compared with those of Fig. 7.5.

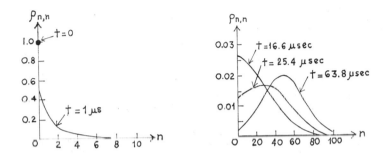

Figure 7.7 Transient buildup from vacuum for a laser 10% above threshold and $\langle n \rangle = 50$ in the steady state (after M. O. Scully, Ref. 3, p. 619)

With similar parameters, but for a laser 30% above threshold, the rise time falls to $t \sim 20\mu$s. The order of magnitude of the rise time is given by $(A - \omega/Q)^{-1}$, which would be the time constant for exponential growth in the linear approximation. In the above example, therefore, the basic adiabatic approximation (7.1.4) is well justified, except very close to the initial instant.

(b) The intrinsic linewidth

We now consider the off-diagonal equations (7.4.35), in the perturbative approximation (7.4.36); adopting the notation (7.4.41), we get

$$\dot{\varrho}_{n,n+p} = -\frac{B}{8}p^2\varrho_{n,n+p} - \left[A - B\left(n + 1 + \frac{p}{2}\right)\right]\left(n + 1 + \frac{p}{2}\right)\varrho_{n,n+p}$$

$$+ \left[A - B\left(n + \frac{p}{2}\right)\right]\sqrt{n(n+p)}\,\varrho_{n-1,n-1+p} - C\left(n + \frac{p}{2}\right)\varrho_{n,n+p}$$

$$+ C\sqrt{(n+1)(n+1+p)}\,\varrho_{n+1,n+1+p}. \tag{7.6.5}$$

The result (7.6.4) is now extended to

$$\varrho_{n,n+p}(t) = \sum_{j=0}^{\infty} \varphi_j(n, p) \exp\left[-\mu_j^{(p)}t\right], \tag{7.6.6}$$

and it may be shown[11a] that

$$\mu_j^{(p)} > 0, \qquad p \neq 0, \tag{7.6.7}$$

so that all nondiagonal terms decay to zero for large times,

$$\lim_{t \to \infty} \varrho_{n,n+p}(t) = 0, \qquad p \neq 0. \tag{7.6.8}$$

We are interested mainly in the neighborhood of the main diagonal [cf. (7.3.7)], i.e., the region

$$p \ll n. \tag{7.6.9}$$

It is then to be expected, by continuity, that the lowest eigenvalue $\mu_0^{(p)}$ be small [cf. (7.6.3)], and that the corresponding eigenfunction $\varphi_0(n, p)$ be close to the diagonal solution (7.5.16). Making the "Ansatz"

$$\varrho_{n,n+p}(t) = \mathcal{N}_p \left[\prod_{j=0}^{n} \left(\frac{A - Bj}{C} \right) \prod_{l=0}^{n+p} \left(\frac{A - Bl}{C} \right) \right]^{\frac{1}{2}} \exp\left[-\mu_0^{(p)}t \right] \quad (p \ll n), \tag{7.6.10}$$

substituting in (7.6.5) and neglecting higher powers of p/n, as well as terms in $\langle n \rangle^{-2}$, one finds[13] that (7.6.5) is satisfied, to a very good approximation, with

$$\mu_0^{(p)} = \tfrac{1}{2}Dp^2, \tag{7.6.11}$$

where

$$D = \frac{C}{2\langle n \rangle} = \frac{\omega}{2Q\langle n \rangle}. \tag{7.6.12}$$

Going back from the interaction picture to the Schrödinger picture [cf. (7.4.2), (7.4.5)], we find

$$\varrho_{n,n+p}^{(S)} = \langle n | e^{-iH_0 t} \varrho^{(I)} e^{iH_0 t} | n + p \rangle$$

$$= \exp(ip\omega t) \varrho_{n,n+p}^{(I)}, \tag{7.6.13}$$

so that (7.6.10) and (7.6.11) yield

$$\varrho_{n,n+p}^{(S)}(t) = \exp(ip\omega t - \tfrac{1}{2}p^2 Dt) \varrho_{n,n+p}^{(S)}(0). \tag{7.6.14}$$

In particular we may apply this result, with $p = 1$, to the expectation value (7.3.7) of the electric field:

$$\langle E(z, t) \rangle = \mathscr{E} \sin(kz) \sum_n \sqrt{n + 1} \left[e^{i\omega t - Dt/2} \varrho_{n,n+1}(0) + c.c. \right]$$

$$= \langle E(z, 0) \rangle e^{-Dt/2} \cos(\omega t). \tag{7.6.15}$$

The line shape is obtained by taking the Fourier transform (in the rotating-wave approximation) of (7.6.15):

$$E(\omega') = E_0 \int_0^\infty \exp\left[-i(\omega' - \omega)t - \frac{D}{2}t\right] dt = \frac{E_0}{i(\omega' - \omega) + (D/2)},$$

$$(7.6.16)$$

which corresponds to a Lorentzian spectrum,

$$|E(\omega')|^2 = \frac{E_0^2}{(\omega' - \omega)^2 + (D/2)^2}, \qquad (7.6.17)$$

with the *intrinsic linewidth* D given by (7.6.12).

According to (5.2.6),

$$\omega/Q - \Delta\Omega \qquad (7.6.18)$$

is the "cavity" mode width in the absence of active atoms. We may rewrite (7.6.12) as

$$D/\Delta\Omega = 1/(2\langle n\rangle). \qquad (7.6.19)$$

Thus, in the laser, the free mode width is reduced by one-half of the factor

$$\frac{1}{\langle n(\omega)\rangle} = \frac{A_{ab}}{W_{ab}}, \qquad (7.6.20)$$

which represents the ratio [cf. (5.3.2), (5.3.4)] of the spontaneous emission rate to the induced emission rate into the given mode. This brings out the role of spontaneous emission as a noise source.

For a typical laser, $\langle n\rangle \gtrsim 10^6$, so that the intrinsic linewidth is very small, and it decreases the higher one goes up above threshold. It is customary to express the linewidth in terms of the average output power P. The average energy W stored in the laser field is given by $W = \langle n\rangle \hbar\omega$, and P is equal to the average power dissipation, i.e., by (5.2.6),

$$P = W\Delta\Omega = \langle n\rangle \hbar\omega\Delta\Omega, \qquad (7.6.21)$$

so that (7.6.19) becomes

$$D = \frac{\hbar\omega}{2P}(\Delta\Omega)^2. \qquad (7.6.22)$$

For a He-Ne laser with $\lambda = 1.1\ \mu$, $L = 100$ cm and a loss of 2% per pass, operating at an output power of $P = 1$ mw, one finds $D \sim 10^{-3}$ cps. This would correspond to a degree of monochromaticity of one part in 10^{17}!

In practice the linewidth is several orders of magnitude bigger, due to secondary effects such as mechanical vibrations of the mirrors. However, by taking special precautions, linewidths of the order of 1 cps have already been obtained for short periods of time.

(c) Physical interpretation

The procedure by which the spectrum (7.6.17) was obtained requires some discussion. According to (7.6.8), the density operator always becomes diagonal in the steady state, so that $\langle E \rangle = 0$ by (7.3.7); this seems to be inconsistent with (7.6.15).

However, in view of the independence between diagonal and off-diagonal elements of the density matrix (§ 7.4), we can always prepare our ensemble in such a way that initially $\varrho_{n,n+1} \neq 0$, so that $\langle E(z, 0) \rangle \neq 0$ in (7.6.15). How are we to interpret the damping that subsequently takes place, according to (7.6.15)? It is not a damping of the amplitude, which is stabilized above threshold. However, the phase is not stabilized and the damping results from a *diffusion process for the phase* in the ensemble of lasers: if we start with a well-defined phase, so that $\langle E \rangle \neq 0$, after a characteristic time $\Delta t \sim D^{-1}$ (of the order of half an hour in the example given above), the phase will have become completely random due to the effect of incoherent noise, so that $\langle E \rangle \to 0$. This does not mean that the phase of the electric field in any given laser becomes random: the diffusion process refers to the phase distribution in the *ensemble*. The phase will evolve differently in different members of the ensemble, under the influence of thermal and spontaneous emission noise, so that it eventually becomes randomly distributed over the ensemble and the ensemble average of $E(z, t)$ tends to vanish. This is another example like that of § 4.3(d), in which the behavior of each individual member of the ensemble is quite different from the ensemble average.

A model for the phase diffusion within the framework of the semiclassical theory was analyzed by Lamb[14]. In addition to the polarization driving term in (6.2.10), he introduced an external noise source, and he showed that, under the influence of this stochastic perturbation, the phase $\varphi(t)$ of the electric field (6.2.13) satisfies

$$\langle (\Delta\varphi)^2 \rangle = \langle [\varphi(t) - \varphi(0)]^2 \rangle = Dt, \qquad (7.6.23)$$

the characteristic result for diffusion processes (Einstein's equation in the theory of Brownian motion), with diffusion constant D.

It then follows from the Wiener-Khintchine theorem [cf. (1.3.33)] that the spectrum is proportional to the Fourier transform of the temporal coherence function, i.e., of

$$\langle e^{i[\omega t + \varphi(t)]} e^{-i\varphi(0)} \rangle = e^{i\omega t} \langle 1 + i[\varphi(t) - \varphi(0)] - \tfrac{1}{2}[\varphi(t) - \varphi(0)]^2 + \cdots \rangle$$
$$\approx e^{i\omega t} \exp\{-\tfrac{1}{2} \langle [\varphi(t) - \varphi(0)]^2 \rangle\} = e^{i\omega t - Dt/2}, \qquad (7.6.24)$$

so that we recover the result (7.6.17), in which D can now be interpreted as the phase diffusion constant.

The Wiener-Khintchine theorem also allows us to define and compute the spectrum even in the stationary case, when ϱ is diagonal and $\langle E \rangle = 0$, by computing the Fourier transform of $G^{(1,1)}(\mathbf{x}, t; \mathbf{x}, t + \tau)$ [cf. (2.3.32)]. For this purpose, it is sufficient to evaluate

$$\text{Tr} \left[\varrho(0) \, a^+(0) \, a(t) \right]. \tag{7.6.25}$$

However, in order to find the time dependence of $a(t)$, we have to take into account the fact that the electromagnetic field is only one part of the complete system, which includes the pumping and damping reservoirs. Thus, we first have to consider the time evolution of the density operator for the complete system, and then to take the trace over the reservoir variables, in order to determine (7.6.25). When this is done[15], the result is again in agreement with (7.6.17) [cf. also § 7.7(c)].

7.7 OTHER APPROACHES

We now give a brief survey of some other approaches to the quantum theory of the laser, which help to shed light on different aspects of the problem.

(a) The Fokker-Planck equation

Although we chose to discuss the density operator equation (7.4.34) in the Fock representation, it is very instructive also to investigate it in the coherent-state representation. For this purpose we employ the diagonal representation (3.4.2) of the density operator,

$$\varrho(t) = \int \varphi(v, t) \, |v\rangle \, \langle v| \, d^2v, \tag{7.7.1}$$

where $|v\rangle$ is a coherent state of the single (resonant) mode present in the laser field.

Substituting (7.7.1) in (7.4.34), we are led to an equation of motion for the (time-dependent) weight function $\varphi(v, t)$. To find the form of this equation, it is convenient to make use of an expression for the weight function due to Lax and Louisell[16].

Let us assume that ϱ has an *antinormally-ordered* power series expansion in terms of a, a^+ [cf. (3.4.5)],

$$\varrho = \varrho_A = \sum_{m,n} r_{m,n} \, a^m (a^+)^n, \tag{7.7.2}$$

where the subscript A indicates antinormal ordering. Inserting the resolution of unity (3.2.31), we find

$$\varrho = \varrho_A = \frac{1}{\pi} \sum_{m,n} r_{m,n} \int a^m \, |v\rangle \, \langle v| \, (a^+)^n \, d^2v$$
$$= \frac{1}{\pi} \int \varrho_A(v, v^*) \, |v\rangle \, \langle v| \, d^2v, \tag{7.7.3}$$

where
$$\varrho_A(v, v^*) = \sum_{m,n} r_{m,n}\, v^m (v^*)^n \tag{7.7.4}$$

is the *numerical function* obtained by substituting a by v and a^+ by v^* in the antinormally-ordered expansion (7.7.2) of the density operator.

Comparing (7.7.1) and (7.7.3), we conclude that

$$\varphi(v, t) = \frac{1}{\pi}\, \varrho_A(v, v^*, t). \tag{7.7.5}$$

The possibly singular character of the weight function φ has its counterpart in possible singularities[17] of the coefficients $r_{m,n}$ in the antinormally-ordered expansion (7.7.2).

In order to apply the above prescription to the equation of motion (7.4.34), we first rewrite the equation in such a way that every term is antinormally ordered[18]. Using for ϱ the antinormally-ordered form ϱ_A, we have to move every a operator to the left and every a^+ to the right in the coefficients. For this purpose we need the commutators between these operators and ϱ_A. Rewriting (3.1.13) in the form

$$[a, (a^+)^n] = n(a^+)^{n-1} = \frac{\partial}{\partial a^+}\, [(a^+)^n], \tag{7.7.6}$$

and similarly for the hermitian conjugate of this relation, we find from (7.7.2)

$$[a, \varrho_A] = \frac{\partial \varrho_A}{\partial a^+}, \qquad [a^+, \varrho_A] = -\frac{\partial \varrho_A}{\partial a}. \tag{7.7.7}$$

We illustrate the method, for simplicity, only in the linear approximation, setting $B = 0$ in (7.4.34). With the help of (7.7.7), we then find the antinormally-ordered form of (7.4.34) in the linear approximation,

$$\frac{\partial \varrho_A}{\partial t} = -\left(\frac{A-C}{2}\right)\left[\frac{\partial}{\partial a}\,(a\varrho_A) + \frac{\partial}{\partial a^+}\,(\varrho_A a^+)\right] + A\,\frac{\partial^2 \varrho_A}{\partial a^+ \partial a}. \tag{7.7.8}$$

Since every term here is in antinormally-ordered form, the solution of this equation of motion also remains antinormally ordered for all times. With the help of (7.7.5), we may immediately transform (7.7.8) into an equation of motion for the weight function $\varphi(v, t)$. We adopt the notation

$$\varphi(v, t) = P(v, v^*, t). \tag{7.7.9}$$

It then follows from (7.7.5) and (7.7.8) that

$$\frac{\partial P}{\partial t} = -\left(\frac{A-C}{2}\right)\left[\frac{\partial}{\partial v}\,(vP) + \frac{\partial}{\partial v^*}\,(v^*P)\right] + A\,\frac{\partial^2 P}{\partial v \partial v^*}. \tag{7.7.10}$$

Let us express this result in terms of real variables, by setting

$$v = x + iy, \tag{7.7.11}$$

and going over from (v, v^*) to (x, y), so that

$$\frac{\partial}{\partial v} = \frac{1}{2}\left(\frac{\partial}{\partial x} - i\frac{\partial}{\partial y}\right). \tag{7.7.12}$$

The result is

$$\frac{\partial P}{\partial t} = -\left(\frac{A - C}{2}\right)\left[\frac{\partial}{\partial x}(xP) + \frac{\partial}{\partial y}(yP)\right] + \frac{A}{4}\left(\frac{\partial^2 P}{\partial x^2} + \frac{\partial^2 P}{\partial y^2}\right)$$

$$= \gamma \operatorname{div}(\mathbf{r}P) + \delta\Delta P, \tag{7.7.13}$$

where $\mathbf{r}$ is the two-dimensional vector with components (x, y) and the parameters γ and δ are defined by (7.7.13).

Equation (7.7.13) is a two-dimensional form of the *Fokker-Planck equation*[19]. In the classical theory of stochastic processes, $P(\mathbf{r}, t)$ would be a probability density. To get some insight into the physical interpretation of (7.7.13), note that, for $\delta = 0$, it reduces to the continuity equation, with a probability current $\mathbf{j} = -\gamma\mathbf{r}P$, so that $-\gamma\mathbf{r}$ plays the role of local flow velocity; for $\gamma = 0$, it reduces to the diffusion equation, with δ playing the role of diffusion constant. Thus, (7.7.13) describes the diffusion of P in the $\mathbf{r}$ plane, in the presence of an external force that gives rise to a convection current[20] with velocity $-\gamma\mathbf{r}$. The term in γ is known as the *drift term*, and that in δ as the *diffusion term*. As we know that the linear approximation only makes sense below threshold, i.e., for $A < C$, we assume that $\gamma > 0$ in (7.7.13). Note that γ then represents the *effective loss* parameter (loss − gain; [cf. (7.5.18)]).

Under these conditions the *fundamental solution* of (7.7.13), i.e., the solution satisfying the initial condition

$$P(\mathbf{r}, 0) = \delta(\mathbf{r} - \mathbf{r}_0) \tag{7.7.14}$$

is given by[21]

$$P(\mathbf{r}, t; \mathbf{r}_0) = \frac{1}{2\pi\sigma^2(t)}\exp\left[-\frac{|\mathbf{r} - \mathbf{r}_0 e^{-\gamma t}|^2}{2\sigma^2(t)}\right], \tag{7.7.15}$$

where

$$\sigma^2(t) = \delta(1 - e^{-2\gamma t})/\gamma. \tag{7.7.16}$$

This satisfies the normalization condition (3.4.71) and is positive definite, so that we may interpret P as a true probability distribution in the $\mathbf{r}$-plane.

The initial condition (7.7.14) corresponds [cf. (7.7.9)] to $\varphi(v, 0) = \delta(v - v_0)$ i.e., to starting with the field in a pure coherent state $|v_0\rangle$. Thereafter, ac-

11*

cording to (7.7.15), the quasiprobability distribution $\varphi(v, t)$ [cf. § 3.4(e)] broadens into a Gaussian distribution with steadily increasing variance [given by (7.7.16)] and centered at the point $v_0 \exp(-\gamma t)$, which approaches the origin as $t \to \infty$. The effective loss γ is also the damping constant for the amplitude of the initial coherent state.

We have to remember [cf. § 7.4(a)] that the above solution of the equations of motion refers to the density operator in the interaction picture. If we transform back to the Schrödinger picture [cf. (7.4.2), (7.4.5)], the effect is to replace v_0 by $v_0 \exp(-i\omega t)$, so that the center of the Gaussian distribution (7.7.15) is now at $v_0 \exp(-i\omega t - \gamma t)$, and it describes a spiral with exponentially decreasing radius about the origin, corresponding to the classical path of a damped two-dimensional harmonic oscillator. The result also corresponds to the classical Brownian motion of a simple harmonic oscillator[22].

The steady-state distribution, which is approached as $t \to \infty$, corresponds to that for thermal light [cf. (4.1.13)], with [cf. (7.7.13)]

$$\langle n \rangle = 2\sigma^2(\infty) = 2\delta/\gamma = A/(C - A). \tag{7.7.17}$$

Comparing this with (4.1.8) and (7.5.10), we see that the result confirms the solution for the laser field below threshold already found in § 7.5.

Above threshold, i.e., for $\gamma < 0$, the nonlinear terms in the equation of motion for the density operator have to be taken into account. This leads to extensions of the Fokker-Planck equation (7.7.10), that have been derived by Lax and Louisell[23] and by Haken and collaborators.[24]

The simplest nonlinear approximation, as suggested by (7.5.15), (6.7.10) and (3.1.25), would be to decrease the effective gain in (7.7.10) by a term proportional to $|v|^2$, in order to represent the nonlinear saturation effect due to induced emission. This would substitute (7.7.13) by an equation of the form

$$\frac{\partial P}{\partial t} = -\operatorname{div}\left[(|\gamma| - \beta r^2)\,\mathbf{r}P\right] + \delta\Delta P, \tag{7.7.18}$$

where β is a saturation parameter. The more elaborate treatments referred to above in fact all lead to an equation of this form when suitable approximations are made (there are some differences in the expressions for the coefficients γ, β, δ in different treatments).

Introducing polar coodinates,

$$\mathbf{r} = (r, \varphi); \quad v = re^{i\varphi}, \tag{7.7.19}$$

the result (7.7.18) becomes

$$\frac{\partial P}{\partial t} = -\frac{1}{r}\frac{\partial}{\partial r}\left[(|\gamma| - \beta r^2)\,r^2 P\right] + \frac{\delta}{r^2}\left[r\frac{\partial}{\partial r}\left(r\frac{\partial P}{\partial r}\right) + \frac{\partial^2 P}{\partial \varphi^2}\right]. \tag{7.7.20}$$

The steady-state solution of this equation (obtained by setting $\partial P/\partial t = 0$) is independent of φ; it is given by

$$\frac{\partial P}{\partial r} = \frac{r}{\delta}(|\gamma| - \beta r^2)\, P, \qquad (7.7.21)$$

i.e., remembering that we set $|\gamma| = -\gamma$ above threshold,

$$P(r) = \mathcal{N} \exp\left(\frac{-\gamma}{2\delta}\, r^2 - \frac{\beta}{4\delta}\, r^4\right), \qquad (7.7.22)$$

where $\mathcal{N}$ is a normalization constant.

Below threshold, i.e., for $\gamma > 0$, this result coincides with the steady-state form of (7.7.15) (Gaussian distribution) if we set $\beta = 0$. Above threshold, we get a distribution peaked at

$$\bar{r}^2 = |\gamma|/\beta = \text{effective gain/saturation}, \qquad (7.7.23)$$

as we see from (7.7.21). This should be compared with (7.5.18). The shape of the quasiprobability distribution $\varphi(v)$ as a function of the normalized variables $x = (\beta/\delta)^{\frac{1}{4}}\, r$ and the "pumping parameter" $p = \gamma/(\beta\delta)^{\frac{1}{2}}$ resembles the curves shown in Fig. 7.5. In the "intensity" variable r^2, we have a Gaussian distribution centered at the point (7.7.23).

Far above threshold, i.e., for $|p| \gg 1$, the variance of this Gaussian distribution becomes very small and we may write

$$r = \bar{r} + \varrho, \qquad (7.7.24)$$

where ϱ represents a small amplitude fluctuation. Separating the variables in (7.7.20) in the form

$$P(r, \varphi, t) = R(\varrho, t)\, \Phi(\varphi, t), \qquad (7.7.25)$$

we find, as a first approximation,

$$\frac{\partial R}{\partial t} = 2\,|\gamma|\, \frac{\partial}{\partial \varrho}\,(\varrho R) + \delta\, \frac{\partial^2 R}{\partial \varrho^2}, \qquad (7.7.26)$$

$$\frac{\partial \Phi}{\partial t} = \frac{\beta\delta}{|\gamma|}\, \frac{\partial^2 \Phi}{\partial \varphi^2}. \qquad (7.7.27)$$

Comparing (7.7.26) with (7.7.13), we see that it is the one-dimensional form of the Fokker-Planck equation, with decay constant $2|\gamma|$; the solution is the one-dimensional form of (7.7.15). Thus, according to (7.7.24), the steady-state amplitude in this approximation consists of two components: a coherent part $\bar{r}$ plus a small fluctuation ϱ, which is Gaussianly distributed, with zero average and variance $(\delta/2|\gamma|)^{\frac{1}{2}}$, which is of the order $|p|^{-1}$ $(\ll 1)$ times smaller than the coherent component.

On the other hand, the phase function Φ obeys the diffusion equation (7.7.27), with diffusion coefficient $\beta\delta/|\gamma| = \delta/\bar{r}^2$ inversely proportional to the output power. This determines the intrinsic linewidth of the laser field [cf. (7.6.19)] and shows that it indeed arises from phase diffusion[25] [cf. § 7.6(c)].

The more elaborate treatments referred to above go considerably further than the qualitative discussion given here. Following a procedure due to Gordon[26], one starts from an equation of motion for the density operator of the laser system, including the atomic variables. One then employs an extension of the diagonal representation (7.7.1) in which some collective atomic variables (the total populations of the atomic levels and the macroscopic dipole moment) are treated in a similar way to the field variables, and this leads to a generalized Fokker-Planck equation. In the adiabatic approximation, the atomic variables can be eliminated, leading to a Fokker-Planck equation for the field variables alone. One can also go beyond the adiabatic approximation, allowing the populations to vary at a rate comparable with that of the field[27]. The effect of detuning (operation at a frequency different from the resonance frequency) on the linewidth has also been determined.

The transient buildup of the laser field from an initial vacuum state has also been investigated[28] by solving the Fokker-Planck equation (7.7.20) with this initial condition. The results look similar to those shown in Fig. 7.7, but they include more realistic situations, in which the average number of photons can be large, as in an actual laser. Experimental results[29] have been found to be in very good agreement with the theory.

(b) The Langevin method

A well-known method to represent the effects due to the interaction with a large reservoir in the theory of Brownian motion is the use of Langevin forces[30]. Thus, for the Brownian motion of a free particle in one dimension, we have the Langevin equation

$$\frac{dv}{dt} + \gamma v = K(t)/m = F(t), \tag{7.7.28}$$

where v is the velocity and the effect of the medium is decomposed into two parts: a systematic part $-\gamma v$, representing friction, and a fluctuating force $K(t) = mF(t)$, with the following properties:

$$\langle F(t) \rangle = 0, \tag{7.7.29}$$

$$\langle F(t)\,F(t') \rangle = 2D\,\delta(t - t'), \tag{7.7.30}$$

so that $F(t)$ is completely random (no correlation between different times). By the Wiener-Khintchine theorem [cf. (1.3.33)], it follows from (7.7.30) that we have a "white noise spectrum". It is assumed, in addition, that the probability distribution for the random variable $F(t)$ is Gaussian, so that higher-order correlation functions factorize in the manner indicated by (4.3.7).

It can then be shown that the probability distribution $P(v, t)$ for $v(t)$ satisfies the Fokker-Planck equation

$$\frac{\partial P}{\partial t} = \gamma \frac{\partial}{\partial v} (vP) + D \frac{\partial^2 P}{\partial v^2}, \qquad (7.7.31)$$

which is the one-dimensional form of (7.7.13). With the above assumptions, we have a stationary Markoff process, i.e., a random process with no "memory" effects, for which knowledge of the probability distribution at the initial time suffices to predict its future behavior (no information on past history is required). This follows specially from (7.7.30).

The Langevin method has been extended to treat quantum noise sources[31]. To explain the procedure, let us go back to the interaction of a single two-level atom with the laser field [§ 7.3(c)], and let us work in the *Heisenberg picture*. In this picture, the equation of motion for the (time-dependent) field operator $a(t)$ is

$$\frac{da}{dt} = i[H, a] = -i\omega a - ig\sigma_-, \qquad (7.7.32)$$

where H is the Hamiltonian (7.3.23) of the atom-field system and we have employed the rotating-wave approximation (7.4.8). For the interaction of the field with the whole active medium of N_0 two-level atoms, it suffices to replace σ_- by $\sum_{j=1}^{N_0} \sigma_-^{(j)}$.

There remains only the interaction between the field and the field loss reservoir (Fig. 6.1). By analogy with (7.7.28), we represent the effect of this interaction by a damping term $-\gamma a$ and a quantum Langevin "force" $F(t)$, so that we finally get

$$\frac{da}{dt} = (-i\omega - \gamma) a - ig \sum_{j=1}^{N_0} \sigma_-^{(j)} + F(t). \qquad (7.7.33)$$

The reservoir average of the fluctuating force $F(t)$ is assumed to vanish, as in (7.7.29). However, $F(t)$ is now an operator, and (7.7.30) is replaced by

$$\langle F(t) F^+(t') \rangle = 2\gamma(\bar{n} + 1) \delta(t - t'), \qquad (7.7.34)$$

$$\langle F^+(t) F(t') \rangle = 2\gamma\bar{n}\delta(t - t'), \qquad (7.7.35)$$

where $\bar{n}$ is the steady-state average occupation number of the field mode in equilibrium with the reservoir; for a reservoir in thermal equilibrium at temperature T, it is given by (4.1.6).

The noncommutative character of the correlations (7.7.34), (7.7.35) is required to preserve in time the commutation relations between a and a^+ [cf. (7.1.5)]. The proportionality between the diffusion coefficients in these relations and the damping constant γ is an example of the "fluctuation-dissipation" relations. The diffusion coefficients and the form of the Langevin forces can be obtained by means of a perturbation calculation (similar to that of § 7.4) of the effect of the system-reservoir interaction. The random Langevin forces are also found to be Gaussian to a good approximation.

In a similar way, one can derive equations of motion for the atomic variables, such as the transition dipole moments σ_- that appear in (7.7.33), with corresponding Langevin forces representing the interaction with the atom reservoirs (Fig. 6.1). By eliminating the atomic variables between the resulting equations and (7.7.33), one is led to an equation of motion for the field variables alone. Finally, this equation can be associated with a Fokker-Planck equation by an extension of the procedure whereby (7.7.31) is associated with (7.7.28). In this way one is led to the quantum versions of the Fokker-Planck equation already discussed in § 7.7(a).

Conversely, one can associate with the quantum Fokker-Planck equation a *classical* Langevin equation. Thus, (7.7.18) is associated with a Langevin equation of the form

$$\dot{v} = (\alpha - \beta |v|^2) v + f(t), \tag{7.7.36}$$

where $\alpha = -\gamma$ represents the effective gain, v is a complex variable, as in (7.7.19), and $f(t)$ is the fluctuating Langevin force. One contribution to $f(t)$, representing the effects of thermal noise and vacuum fluctuations, comes from $F(t)$ in (7.7.33); another contribution comes from the effect of noise sources on the atomic dipole moment terms σ_- in (7.7.33), and it contains the spontaneous emission noise.

The result (7.7.36) should be compared with the Van der Pol equation (6.7.10) of the semiclassical theory. The associated classical system is often referred to as a *rotating-wave Van der Pol oscillator* [because the rotating-wave approximation was employed in the derivation leading to (7.7.36)]. Its properties have been investigated in detail by Lax[32].

(c) The quantum regression theorem

In § 7.6(c), we mentioned that the temporal coherence function (7.6.25), which defines the spectrum via the Wiener-Khintchine theorem, decays in time in the same way as the average electric field (7.6.15) in an ensemble prepared so as to have a nonvanishing initial value for the average field. This justifies the identification of (7.6.17) with the spectrum of the laser field.

This result is an illustration of the validity for the present system of Onsager's "regression hypothesis"[33], according to which the regression of fluctuations from a nonequilibrium state obeys, on the average, the same laws as the corresponding macroscopic irreversible process.

It was shown by Lax[34] that this follows from the assumption that the system and the reservoir are initially uncorrelated, so that the total density operator may be factored at the initial time [cf. (7.4.11), (7.4.23)]. According to Lax's *quantum regression theorem*, if this assumption is made, and if the average of any system operator M at time t can be expressed in terms of the averages of a set of system operators M_j at an earlier time t' by

$$\langle M(t) \rangle = \sum_j O_j(t, t') \langle M_j(t') \rangle \quad (t > t'), \tag{7.7.37}$$

then, for any other system operator N,

$$\langle M(t) N(t') \rangle = \sum_j O_j(t, t') \langle M_j(t') N(t') \rangle, \tag{7.7.38}$$

with the same coefficients $O_j(t, t')$. In this way, a two-time correlation is reduced to a one-time average (mean equation of motion). In particular, (7.6.25) is reduced to (7.3.7), i.e., to (7.6.15). Lax has also extended the result to multitime averages[35], and he has shown that the regression theorem is equivalent to the assumption that the system is Markoffian.

References

1. R. K. Wangsness and F. Bloch, *Phys. Rev.* **89**, 728 (1953).
2. M. O. Scully and W. E. Lamb, *Phys. Rev.* **159**, 208 (1967).
3. M. O. Scully, *QO* (see Ch. 1, Ref. 3), p. 586.
4. Cf. I. R. Senitzky, *Phys. Rev.* **119**, 670 (1960), and W. H. Louisell, *Radiation and Noise in Quantum Electronics*, McGraw-Hill, New York (1964), p. 255.
5. Cf. W. H. Louisell, loc. cit. and *QO*, p. 695; R. J. Glauber, *QO*, p. 32, and the references quoted there.
6. For a more detailed discussion of the reduced density operator, see U. Fano, *Rev. Mod. Phys.* **29**, 74 (1957), and D. Ter Haar, *Rep. Progr. Phys.* **24**, 304 (1961).
7. M. O. Scully, *QO*, p. 586.
8. Except for the substitution $R_{n',n} \to R^*_{n',n}$ when $\Delta \neq 0$.
9. For a discussion of the master equation, cf., e.g., *Fundamental Problems in Statistical Mechanics*, editor E. G. D. Cohen, North-Holland Publishing Co., Amsterdam (1962).
10. M. O. Scully and W. E. Lamb, *Phys. Rev.* **179**, 368 (1969).
11. G. Freed and H. Haus, *Phys. Rev. Letters* **15**, 943 (1965); A. W. Smith and J. A. Armstrong, *Phys. Rev. Letters* **19**, 650 (1966); F. T. Arecchi, *Phys. Lett.* **25A**, 59 (1967); F. Davidson and L. Mandel, *Phys. Letters* **25A**, 700 (1967); cf. the survey article by F. T. Arecchi, *QO*, p. 57.

11a. W. E. Lamb and Y. K. Wang, to be published.

12. M. O. Scully, W. E. Lamb, and M. Sargent III, *IV International Conference on Quantum Electronics* (Phoenix, Ariz., 1966); M. O. Scully, *QO*, p. 619.

13. M. O. Scully and W. E. Lamb, *Phys. Rev.* **159**, 208 (1967), Appendix III.

14. W. E. Lamb, *QOE* (see Ch. 1., Ref. 2), p. 377.

15. M. O. Scully, *QO*, p. 625.

16. M. Lax and W. H. Louisell, *J. Quantum Electron.* **QE 3**, 47 (1967).

17. K. E. Cahill and R. J. Glauber, *Phys. Rev.* **177**, 1857, 1882 (1969).

18. R. Bonifacio and F. Haake, *Z. Physik* **200**, 526 (1967).

19. Cf., e.g., M. C. Wang and G. E. Uhlenbeck, *Rev. Mod. Phys.* **17**, 323 (1945); P. Lévy, *Processus Stochastiques et Mouvement Brownien*, Gauthier-Villars, Paris (1965).

20. Cf. R. Fürth, in *Differentialgleichungen der Physik*, ed. by P. Frank and R. v. Mises, Friedr. Vieweg & Sohn, Braunschweig (1935), vol. 2, p. 593.

21. S. Chandrasekhar, *Rev. Mod. Phys.* **15**, 1 (1943), p. 34.

22. M. C. Wang and G. E. Uhlenbeck, loc. cit.; R. J. Glauber, *QO*, pp. 32–46.

23. M. Lax and W. H. Louisell, *J. Quantum Electron.* **QE 3**, 47 (1967); W. H. Louisell, *QO*, p. 680 (and the references quoted there).

24. H. Haken, H. Risken, and W. Weidlich, *Z. Physik* **204**, 223 (1967); H. Haken and W. Weidlich, *QO*, p. 630 (and the references quoted there); H. Haken, *Laser Theory*, *Handb. der Phys.*, Vol. XXV/2c, Springer-Verlag, Berlin (1970).

25. H. Risken, *Z. Physik* **186**, 85 (1965); F. T. Arecchi, *QO*, p. 82.

26. J. P. Gordon, *Phys. Rev.* **161**, 367 (1967).

27. M. Lax, *Phys. Rev.* **157**, 213 (1967); M. Lax and H. Yuen, *Phys. Rev.* **172**, 362 (1968).

28. H. Risken and H. D. Vollmer, *Z. Physik* **204**, 240 (1967).

29. D. Meltzer, Ph. D. thesis, University of Rochester (1970); D. Meltzer and L. Mandel, *Phys. Rev. Letters* **25**, 1151 (1970), *Phys. Rev. A* **3**, 1763 (1971).

30. Cf., e.g., M. C. Wang and G. E. Uhlenbeck, loc. cit.

31. I. R. Senitzky, *Phys. Rev.* **119**, 670 (1960); **123**, 1525 (1961); M. Lax, *Phys. Rev.* **145**, 110 (1966).

32. M. Lax, *Phys. Rev.* **160**, 290 (1967); R. D. Hempstead and M. Lax, *Phys. Rev.* **161**, 350 (1967).

33. L. Onsager, *Phys. Rev.* **37**, 405 (1931); **38**, 2265 (1931); L. Onsager and S. Machlup, *Phys. Rev.* **91**, 1505 (1953); H. B. G. Casimir, *Rev. Mod. Phys.* **17**, 343 (1945).

34. M. Lax, *Phys. Rev.* **129**, 2342 (1963); **157**, 213 (1967).

35. M. Lax, *Phys. Rev.* **172**, 350 (1968).

CHAPTER 8

Superradiance, Photon Echoes and Self-induced Transparency

8.1 INTRODUCTION

THE PHENOMENA THAT we are now going to treat are collective effects, involving the emission or absorption of light *coherently* by a large number of atoms. However, this type of coherence is quite different from that discussed up to now: it does not refer to the statistical properties of the *light* field that interacts with the atoms, but rather to the fact that the entire set of *atoms* behaves collectively in a coherent way, like a single quantum system.

This *coherence among the atoms* means that the radiation rate from a system of N atoms becomes proportional to N^2 rather than N, as it would be for ordinary incoherent emission. The system can then emit a very intense and very short light pulse. This effect is known as *superradiance*[1].

In contrast with a laser, no resonant cavity (i.e., mirrors) is needed to provide feedback. On the other hand, the conditions for the excitation of a superradiant state are very difficult to realize in practice, and superradiant effects do not seem to play a significant role in the laser. These effects, like photon bunching and the Hanbury Brown and Twiss effect (§ 4.3), are related with the clustering tendency of bosons.

The experimental observation of superradiance has been limited thus far to the phenomenon of *photon echoes*. Under certain conditions, a resonant medium, excited successively by two coherent intense and short pulses, can become superradiant and spontaneously produce a third pulse (echo)[2].

Another very surprising effect that may take place in the propagation of intense short pulses in a resonant medium is *self-induced transparency*: for a sufficiently intense pulse of suitable shape, the medium behaves as if it were completely transparent, in spite of the fact that it would strongly absorb pulses of weaker intensity[3].

Characteristically, these nonlinear effects are associated with the propagation of very short pulses, of duration much shorter than the lifetime associated with damping processes (homogeneous broadening), but possibly comparable with lifetimes associated with inhomogeneous line broadening (cf. § 5.3).

8.2 PROPERTIES OF TWO-LEVEL SYSTEMS

The model for the medium will be an assembly of two-level atoms, just like the model for the active medium in a laser. Again we will make the fundamental assumption (iii) of § 6.1: The only interaction among the atoms arises from their common coupling with the radiation field. The effects of this indirect interaction are far from negligible, however: we will see that the emission of radiation from an atom can be drastically affected by the presence of other atoms, even though it does not interact directly with them (cf. § 8.4).

The analogy between two-level atoms and spins that was pointed out in § 6.3(b) plays a basic role in all that follows. The zero of energy is taken halfway between the two levels (Fig. 6.3) so that

$$E_a = -E_b = \hbar\omega_0/2. \tag{8.2.1}$$

For a single atom interacting with the field, in the electric-dipole approximation, the Hamiltonian is [cf. (6.3.9)]

$$\mathscr{H} = \mathscr{H}_A + \hbar V = \mathscr{H}_A - \mathbf{p}\cdot\mathbf{E}. \tag{8.2.2}$$

In contrast with the case of the laser, in which we assumed the electric field to be linearly polarized, it is convenient here to describe the field in terms of circularly polarized components. Accordingly, taking the z-axis along the direction of propagation, the interaction term in (8.2.2) becomes

$$\hbar V = -\tfrac{1}{2}(p^+E_- + p^-E_+), \tag{8.2.3}$$

where

$$p^\pm = p_x \pm ip_y = e(x \pm iy); \quad E_\pm = E_x \pm iE_y. \tag{8.2.4}$$

The operators $p^\pm$ have nonvanishing matrix elements between states with $\Delta l = 1, \Delta m = \mp 1$, respectively, where m is the magnetic quantum number. For definiteness, let us assume that level a has quantum numbers $(l + 1, m + 1)$, and b has quantum numbers (l, m). Then, with a suitable choice of phases for the stationary eigenfunctions ψ_a, ψ_b, the matrix element

$$p_{ab}^+ = \int \psi_a^*(\mathbf{x})\, p^+\psi_b(\mathbf{x})\, d^3x = 2p \tag{8.2.5}$$

can be taken as real, and (8.2.3) yields

$$\hbar V_{ab} = -\tfrac{1}{2}p_{ab}^+E_- = -pE_-,$$

$$\hbar V_{ba} = -\tfrac{1}{2}p_{ba}^-E_+ = -pE_+. \tag{8.2.6}$$

In (8.2.5), $p = |p_x| = |p_y|$ is the common magnitude of the transition dipole moment components[4].

Comparing (8.2.6) with (6.3.34), we see that

$$V_1 = -\varkappa E_x, \quad V_2 = -\varkappa E_y,$$ (8.2.7)

where

$$\varkappa = 2p/\hbar.$$ (8.2.8)

It is convenient to use, instead of the Pauli spin matrices σ_i of § 6.3(b), the analogues of the spin operators R_i,

$$R_i = \tfrac{1}{2}\sigma_i \quad (i = 1, 2, 3),$$ (8.2.9)

which [cf. (6.3.36)] obey the normal angular momentum commutation rules,

$$[R_1, R_2] = iR_3 \quad \text{(and circular permutations).}$$ (8.2.10)

In terms of these operators, (6.3.45) becomes

$$\langle \mathbf{R} \rangle = \tfrac{1}{2}\mathbf{r},$$ (8.2.11)

and (6.3.46), (6.3.43) (taking into account (8.2.7)) may be rewritten as

$$\mathscr{H} = \hbar\boldsymbol{\omega} \cdot \mathbf{R} = -\mathscr{P} \cdot \mathscr{E},$$ (8.2.12)

where the "pseudo-electric dipole moment" operator $\mathscr{P}$ is defined by

$$\mathscr{P} = 2p\mathbf{R},$$ (8.2.13)

and the "pseudo-electric field" $\mathscr{E}$ is defined by

$$\mathscr{E} = E_x\hat{\mathbf{x}} + E_y\hat{\mathbf{y}} - \frac{\omega_0}{\varkappa}\hat{\mathbf{z}}.$$ (8.2.14)

For a superposition state, $\psi(\mathbf{x}) = a\psi_a(\mathbf{x}) + b\psi_b(\mathbf{x})$, the expectation value of $\mathbf{p}$ follows from (8.2.5), (6.3.29), (6.3.30) and (8.2.11):

$$\langle \mathbf{p} \rangle = \operatorname{Re} \mathbf{p}_{ab} \, r_1 - \operatorname{Im} \mathbf{p}_{ab} \, r_2 = p(r_1\mathbf{x} + r_2\hat{\mathbf{y}})$$
$$= \langle \mathscr{P}_x\hat{\mathbf{x}} + \mathscr{P}_y\hat{\mathbf{y}} \rangle.$$ (8.2.15)

Thus, the expectation values of the first two components of $\mathscr{P}$ are related to real and imaginary parts of the transition dipole moment, respectively, and the first two components of $\mathscr{E}$ are those of the electric field. This explains the names given to these expressions.

Finally, the equation of motion (6.3.42) becomes

$$\frac{d}{dt} \langle \mathscr{P} \rangle = \varkappa \langle \mathscr{P} \rangle \times \mathscr{E},$$ (8.2.16)

and it may be thought of as describing the precession of the pseudo-electric dipole moment around the pseudo-electric field.

8.3   THE COOPERATION NUMBER

Let us now consider a system of N two-level atoms, which may be treated as a gas (cf. § 5.3), although it need not actually be a gas: the active medium in a solid-state laser constitutes only a small fraction of the total material, embedded in a host medium. The average spacing between active atoms is assumed to be sufficiently large that there is no appreciable overlap between the wave functions of neighboring atoms; it follows that we need not worry about symmetrization problems of the total wave function of the system. We will also neglect the effects of atomic motion, treating the atoms as if they were at rest.

In view of the basic assumption that there is no direct interaction among the atoms, the hamiltonian for the whole system interacting with the radiation field can be written in the form

$$\mathscr{H} = \sum_{j=1}^{N} \mathscr{H}_A^{(j)} - \sum_{j=1}^{N} \mathbf{p}^{(j)} \cdot \mathbf{E}(\mathbf{x}_j) = \mathscr{H}_0 + \mathscr{H}_I$$

$$= \hbar\omega_0 \sum_{j=1}^{N} R_3^{(j)} - 2p \sum_{j=1}^{N} [R_1^{(j)} E_x(\mathbf{x}_j) + R_2^{(j)} E_y(\mathbf{x}_j)], \qquad (8.3.1)$$

where $\mathbf{E}(\mathbf{x}_j)$ is the electric field at the position of the jth atom, and we are using the notations of (8.2.12)–(8.2.14), with the superscript (j) associated with the jth atom. Note that the $R_i^{(j)}$ operators for different atoms commute:

$$[R_i^{(j)}, R_k^{(j')}] = 0 \quad (j \neq j'). \qquad (8.3.2)$$

For simplicity we assume, initially, that all the atoms are contained within a volume of dimensions much smaller than the wavelength of the radiation field. The effect of removing this assumption will be discussed later. It then follows that all atoms see essentially the same field, i.e., $\mathbf{E}$ does not depend on $\mathbf{x}_j$ in (8.3.1), which then becomes

$$\mathscr{H} = \hbar\omega_0 \mathscr{R}_3 - 2p(E_x \mathscr{R}_1 + E_y \mathscr{R}_2)$$

$$= -2p \mathscr{R} \cdot \mathscr{E} = -\mathscr{P} \cdot \mathscr{E}, \qquad (8.3.3)$$

where $\mathscr{E}$ is still defined by (8.2.14), but now, instead of (8.2.9), we have

$$\mathscr{R}_i = \sum_{j=1}^{N} R_i^{(j)} \quad (i = 1, 2, 3). \qquad (8.3.4)$$

The rate of spontaneous emission of radiation from the system in a transition from an initial state $|\psi_i\rangle$ to a final state $|\psi_f\rangle$ is proportional to [cf. (8.3.1), (8.3.3)]

$$|\langle\psi_f| \mathscr{H}_I |\psi_i\rangle|^2 = 4p^2 |\langle\psi_f| (E_x \mathscr{R}_1 + E_y \mathscr{R}_2) |\psi_i\rangle|^2. \qquad (8.3.5)$$

How does this radiation rate depend on the choice of the initial state $|\psi_i\rangle$, i.e., on the way in which the system is prepared?

The conventional choice for the states $|\psi\rangle$ of the atomic system is a direct product of single-atom states,

$$|\psi\rangle = \prod_{j=1}^{N} |\psi_j\rangle. \qquad (8.3.6)$$

A typical state of this kind can be written as

$$|\psi\rangle = (+ + - + - - \cdots + -), \qquad (8.3.7)$$

where each $+$ $(-)$ stands for an atom in the upper (lower) energy state a (b). The dependence on atomic coordinates is ignored (atom j was assumed to be at rest in the fixed position $\mathbf{x}_j$). The state (8.3.7) is an eigenstate of the unperturbed Hamiltonian, with

$$\mathscr{H}_0 |\psi\rangle = \hbar\omega_0 \mathscr{R}_3 |\psi\rangle = m\hbar\omega_0 |\psi\rangle, \qquad (8.3.8)$$

where

$$m = \tfrac{1}{2}(N_+ - N_-), \qquad (8.3.9)$$

N_+ (N_-) denoting the number of $+$ $(-)$ signs in (8.3.7). We have

$$N = N_+ + N_-. \qquad (8.3.10)$$

The unperturbed energy of the state (8.3.7) depends only on m, and not on the way in which the $+$ and $-$ are distributed among the atoms, so that an eigenstate of this type with energy $m\hbar\omega_0$ has a degeneracy

$$d_m = \frac{N!}{(N_+)!(N_-)!} = \frac{N!}{(N/2 + m)!(N/2 - m)!}. \qquad (8.3.11)$$

This degeneracy is largest for $m = 0$, i.e., when exactly half the atoms are excited, and it then becomes quite large for large N.

It follows from (8.3.4), (8.3.2) and (8.2.10) that the operators $\mathscr{R}_i$ for the N-atom system still obey the commutation rules for angular momentum operators,

$$[\mathscr{R}_1, \mathscr{R}_2] = i\mathscr{R}_3 \quad \text{(and circular permutations)}, \qquad (8.3.12)$$

In particular,

$$[\mathscr{R}_3, \mathscr{R}_\pm] = \pm\mathscr{R}_\pm, \quad \mathscr{R}_\pm = \mathscr{R}_1 \pm i\mathscr{R}_2, \qquad (8.3.13)$$

so that $\mathscr{R}_\pm$ correspond to the raising and the lowering operator, respectively. According to (8.3.5), therefore, we have the selection rule

$$\Delta m = \pm 1. \qquad (8.3.14)$$

However, in view of the above-mentioned degeneracy, there are in general a large number of possible final states, compatible with this selection rule, which are accessible from an initial state of type (8.3.7).

The starting point of Dicke's theory of superradiance[1] is the observation that the degeneracy can be reduced, by choosing a new set of stationary states, in such a way that the transition matrix element in (8.3.5) connects the initial state with at most one state of higher energy and one of lower energy. For this purpose, he used the analogy with spin $\frac{1}{2}$ systems to define the analogue of the "total spin".

The operator

$$\mathscr{R}^2 = \mathscr{R}_1^2 + \mathscr{R}_2^2 + \mathscr{R}_3^2 \tag{8.3.15}$$

commutes with the unperturbed hamiltonian (8.3.8). Therefore, we can construct simultaneous eigenstates $|r, m\rangle$ of $\mathscr{R}^2$ and $\mathscr{H}_0$, with

$$\mathscr{R}^2 |r, m\rangle = r(r + 1) |r, m\rangle, \tag{8.3.16}$$

$$(\hbar\omega_0)^{-1}\mathscr{H}_0 |r, m\rangle = \mathscr{R}_3 |r, m\rangle = m |r, m\rangle. \tag{8.3.17}$$

It follows from (8.3.12), (8.3.15), as in the usual theory of angular momentum, that r may be an integer or half-integer, and m, for a given r, varies from $-r$ to r; by (8.3.9) and (8.3.10), the maximum value of r is $N/2$; thus,

$$|m| \leqq r \leqq N/2. \tag{8.3.18}$$

The quantum number r was called by Dicke the *cooperation number*. The reason for this name is that, as will be seen below, r characterizes the coherence among the atoms mentioned in § 8.1.

The complete set of states $|r, m\rangle$ may be constructed by the standard angular momentum procedure, making use of the lowering operator $\mathscr{R}_-$ [cf. (8.3.13)], starting from the nondegenerate state $|N/2, N/2\rangle = (+ + + ... +)$. Applying $\mathscr{R}_-^s$ ($s = 1, 2, ..., N$) to this state (and normalizing), we generate the remaining eigenstates with the same $r = N/2$, and $m = (N/2) - 1, ..., -N/2$. They are all nondegenerate, i.e., each state $|N/2, m\rangle$ has degeneracy $D_0 = d_{N/2} = 1$ [cf. (8.3.11)].

There are $d_{(N/2)-1} = N$ states with $m = (N/2) - 1$ [cf. (8.3.11)]. One is $|N/2, (N/2) - 1\rangle$; the remaining ones must all have $r = (N/2) - 1$, by (8.3.18), and they can be constructed by the standard orthonormalization procedure. The degeneracy of the state $|(N/2) - 1, (N/2) - 1\rangle$ is therefore $D_1 = d_{(N/2)-1} - D_0 = d_{(N/2)-1} - d_{N/2} = N - 1$. All other states $|(N/2) - 1, m\rangle$ can again be constructed by applying the lowering operator to this state, and each has the same degeneracy.

Continuing in this fashion, we generate the complete set of states $|r, m\rangle$, and we find that each state $|r, m\rangle$ has a degeneracy

$$D_{(N/2)-r} = d_r - d_{r+1} = \frac{(2r + 1) N!}{\left(\dfrac{N}{2} + r + 1\right)!\left(\dfrac{N}{2} - r\right)!}. \tag{8.3.19}$$

8.4 SUPERRADIANCE

(a) Superradiant states

According to (8.3.15) and (8.3.3), the operator $\mathscr{R}^2$ commutes also with the interaction hamiltonian, so that the selection rule (8.3.14) is extended to

$$\Delta r = 0, \qquad \Delta m = \pm 1. \tag{8.4.1}$$

In view of the construction for the states $|r, m\rangle$ outlined in § 8.3, it follows that, if we choose $|\psi_i\rangle$ as a given state $|r, m\rangle$ in (8.3.5), the transition rate is nonvanishing only for the two neighboring states, $|r, m \pm 1\rangle$. In particular, the spontaneous emission rate $I(r, m)$ from the state $|r, m\rangle$ is proportional to the matrix element[5]

$$|\langle r, m - 1| \mathscr{R}_- |r, m\rangle|^2 = (r + m)(r - m + 1),$$

i.e.,

$$I(r, m) = (r + m)(r - m + 1) I_0. \tag{8.4.2}$$

In particular, for $r = m = \frac{1}{2}$, we find $I = I_0$, so that I_0 is the spontaneous emission rate for a *single excited atom*.

How does the radiation rate (8.4.2) depend on (r, m)? If all N atoms are initially excited, $|\psi_i\rangle = |N/2, N/2\rangle$, we get

$$I(N/2, N/2) = NI_0, \tag{8.4.3}$$

which is precisely the result we would expect for N atoms radiating independently (incoherently). We have $\partial I(r, m)/\partial m = -2m + 1 = 0$ for $m = 1/2$, so that assuming, e.g., that N is even, $I(r, m)$ is maximum when $m = 0$ or $m = 1$, i.e., if r is is large, when about half the atoms are excited: $I(r, 0) = I(r, 1) = r(r + 1) I_0$. This increases with r, so that the absolute maximum radiation rate is

$$I(N/2, 0) = \frac{N}{2}\left(\frac{N}{2} + 1\right) I_0 \; (\approx (N^2/4) I_0 \text{ for large } N). \tag{8.4.4}$$

For large N, we see that the state $|N/2, 0\rangle$ radiates as if all excited atoms were "in phase"; the radiation rate is proportional to N^2, rather than N, as mentioned in § 8.1. This state is known as a *superradiant state*. Clearly, superradiance is achieved for large values of the cooperation number r and small values of $|m|$.

If the initial state is taken to be a superposition of the states $|r, m\rangle$, then, because of the orthonormality of these states, the corresponding radiation rate is

$$I = \sum_{r,m} P_{r,m} I(r, m), \tag{8.4.5}$$

where $P_{r,m}$ is the probability of being in the state $|r, m\rangle$. Thus, the radiation rate for any superposition state is smaller than (8.4.4).

The state with $r = N/2, m = -(N/2) + 1$ corresponds to having only one atom excited, but the corresponding radiation rate is

$$I(N/2, -(N/2) + 1) = NI_0, \tag{8.4.6}$$

i.e., it is N times the single-atom rate! Thus, in this case, we also have coherence among the atoms (but not superradiance): we do not know which atom is excited; the excitation is distributed symmetrically among all the atoms. For $r = m = 0$ (which is only possible for an even number of atoms), we get $I = 0$: this state does not radiate at all; it may be compared with a classical set of dipoles oscillating in pairs with opposite phases.

The simplest illustration of the above results corresponds to a two-atom system. The actual expressions for the states $|r, m\rangle$ for $N = 2$ in terms of the states (8.3.7) are shown in Fig. 8.1, together with the corresponding

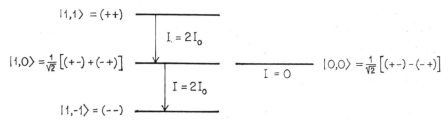

Figure 8.1   The states $|r, m\rangle$ for $N = 2$ and the corresponding transition rates.

transition rates (8.4.2). The states $r = 1$ are analogous to the triplet state for two spin-$\frac{1}{2}$ particles, and $r = 0$ is analogous to the singlet case. The singlet-triplet transition is forbidden, so that $|0, 0\rangle$ is a radiationless state. On the other hand, the radiation rate for $|1, 0\rangle$ is $2I_0$, as in (8.4.6). If we prepare the system initially in the state $(+ -)$, in which only atom No. 1 is excited, it corresponds to $(+ -) = (1/\sqrt{2})(|1, 0\rangle + |0, 0\rangle)$, so we have equal probabilities for the system to be in the triplet and in the singlet state. By (8.4.5), the initial radiation rate is $I = \frac{1}{2} \cdot 2I_0 = I_0$, the same as for an isolated atom, but, after a very long time, the probability that a photon has not been emitted approaches $\frac{1}{2}$, rather than zero: there is still a probability $\frac{1}{2}$ for the energy to be trapped in the radiationless "singlet" mode. We see, therefore, that the presence of another atom in the ground state (within a distance $\ll$ wavelength) has a strong effect on the decay of the first atom, through their common interaction with the radiation field (even though there is no direct interaction between the atoms).

An equivalent way to formulate these results[6] is to note that the hamiltonian (8.3.3) involves all the atoms in a symmetric way, so that it commutes with permutation operators among the atoms. One can then classify the states in terms of eigenvalues of the cyclic permutation operators, and there can be no off-diagonal matrix elements, connecting states with different eigenvalues. An illustration is again provided by the example of Fig. 8.1.

(b) Classical model

If a system is prepared in a state with a definite and very large value of r, we can draw on the analogy between $\mathscr{R}$ and angular momentum to make use of classical limit considerations, and, in particular, of the vector model of angular momentum.

By (8.3.3) and (8.3.16),

$$\langle \mathscr{P}^2 \rangle = 4p^2 \langle \mathscr{R}^2 \rangle = 4p^2 r(r + 1), \tag{8.4.7}$$

where $\mathscr{P}$ is the dipole moment operator for the whole system. The result (8.2.16), which is valid for each component $\mathscr{P}^{(j)}$ of $\mathscr{P} = \sum_j \mathscr{P}^{(j)}$, remains valid for the resultant (because each atom sees the same pseudo-electric field $\mathscr{E}$), so that

$$\frac{d}{dt} \langle \mathscr{P} \rangle = \varkappa \langle \mathscr{P} \rangle \times \mathscr{E}. \tag{8.4.8}$$

For large r, we can interpret (8.4.7) and (8.4.8) in terms of a classical model[7], as describing a spinning top with large angular momentum, carrying an electric dipole moment of fixed magnitude [given by (8.4.7)] along its axis. The "magnetic quantum number" m becomes

$$m = r \cos \theta, \tag{8.4.9}$$

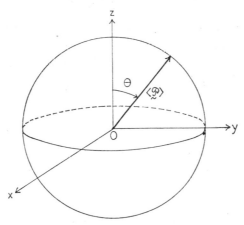

Figure 8.2   The classical model

where θ is the angle between $\langle \mathscr{P} \rangle$ and the z-axis (Fig. 8.2). By (8.3.17), the corresponding internal energy of the gas is

$$\langle \mathscr{H}_0 \rangle = m\hbar\omega_0 = \hbar\omega_0 r \cos\theta. \tag{8.4.10}$$

The spontaneous emission rate (8.4.2) may be approximated by

$$I \approx I_0(r^2 - m^2) = I_0 r^2 \sin^2\theta, \tag{8.4.11}$$

which is maximum for $\theta = \pi/2$ $(m = 0)$, corresponding to a superradiant state. By energy conservation, (8.4.11) must also represent the rate of decrease of the internal energy of the gas, so that, by (8.4.10),

$$I = -\frac{d}{dt}\langle \mathscr{H}_0 \rangle = \hbar\omega_0 r \, \theta \sin\theta = I_0 r^2 \sin^2\theta,$$

i.e.,

$$\theta = \frac{I_0 r}{\hbar\omega_0} \sin\theta = \alpha \sin\theta. \tag{8.4.12}$$

Setting

$$\varphi = 2\theta, \tag{8.4.13}$$

we find that (8.4.12) implies

$$\ddot{\varphi} = \alpha^2 \sin\varphi. \tag{8.4.14}$$

This may be compared with the equation of motion of a simple pendulum of length l,

$$\ddot{\varphi} = (g/l) \sin\varphi. \tag{8.4.15}$$

If we identify α^2 with g/l, the result (8.4.12) corresponds to the energy integral of this equation of motion,

$$E = T + V = \tfrac{1}{2}ml^2\dot{\varphi}^2 + mgl\cos\varphi$$

$$= 2ml^2\alpha^2 \sin^2(\varphi/2) + mgl\cos\varphi = mgl. \tag{8.4.16}$$

Thus, (8.4.12) corresponds to the pendulum motion with total energy E equal to that which it would have if released at rest from its vertical position of unstable equilibrium, $\varphi = 0$ (Fig. 8.2). The solution of (8.4.12) is

$$\int_{\theta_0}^{\theta} \frac{d\theta}{\sin\theta} = (\ln |\tan(\theta/2)|)_{\theta_0}^{\theta} = \alpha t,$$

i.e., assuming that $\theta = \theta_0 = \pi/2$ at $t = 0$,

$$\tan(\theta/2) = e^{\alpha t}, \quad \sin\theta = \text{sech}(\alpha t). \tag{8.4.17}$$

As illustrated in Fig. 8.3, this corresponds to the aperiodic circular pendulum solution[8], in which the pendulum starts from rest at the vertical unstable equilibrium position $(\varphi = \theta = 0)$ at $t \to -\infty$, and goes back to

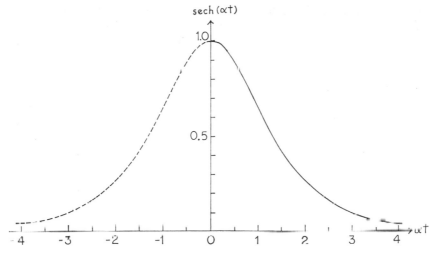

Figure 8.3   The hyperbolic secant solution

this position ($\varphi = 2\pi, \theta = \pi$) as $t \to \infty$ after one complete revolution. The maximum velocity is attained at $t = 0$, $\varphi = \pi$, i.e., $\theta = \pi/2$, corresponding to superradiance [cf. (8.4.11)].

If the system is excited into the superradiant state at $t = 0$, this corresponds to starting the pendulum at $\varphi = \pi$: the vector $\langle \mathscr{P} \rangle$ in Fig. 8.2 "falls" from $\theta = \pi/2$ to $\theta = \pi$; the time behavior for this process is described by the curve in full line in Fig. 8.3.

According to (8.2.15) and (8.3.3), the transition dipole moment for the system corresponds to the "transverse" component of $\langle \mathscr{P} \rangle$,

$$\langle \mathscr{P}_\perp \rangle = \langle \mathscr{P} \rangle - \mathscr{P}_z \hat{\mathbf{z}} ; \quad |\langle \mathscr{P}_\perp \rangle| = |\langle \mathscr{P} \rangle| \sin \theta. \qquad (8.4.18)$$

In addition to the shrinking of $\langle \mathscr{P}_\perp \rangle$ due to the "fall" of $\langle \mathscr{P} \rangle$ (radiation damping), we have to take into account that $\langle \mathscr{P}_\perp \rangle$ rotates around the z-axis with angular velocity ω_0, due to the precession equation (8.4.8), which contains a term

$$\langle \dot{\mathscr{P}}_\perp \rangle = \omega_0 \hat{\mathbf{z}} \times \langle \mathscr{P}_\perp \rangle. \qquad (8.4.19)$$

By (8.4.18) and (8.4.19), the time dependence of the radiated field is given by

$$A(t) = \begin{cases} e^{i\omega_0 t} \sin \theta, & t > 0, \\ 0 & , \quad t < 0, \end{cases} \qquad (8.4.20)$$

where $\sin \theta$ is given by (8.4.17). Asymptotically, for $t \gg \alpha^{-1}$, this yields $A(t) \approx 2 \exp(i\omega_0 t - \alpha t)$, but the decay law in the relevant interval $0 \leq t \lesssim \alpha^{-1}$ is by no means exponential (Fig. 8.3). Correspondingly, the line shape is not Lorentzian [cf. (6.4.22)]. The line width is of the order of the inverse lifetime,

$$\Delta\omega \sim \alpha = r \frac{I_0}{\hbar\omega_0} = r\gamma, \tag{8.4.21}$$

where γ, in view of the interpretation of I_0 [cf. (8.4.2)], is the natural line-width for a single atom.

Thus, the superradiant emission consists of a single pulse, of very high intensity $I \approx r^2 I_0$ ($r \leq N/2$), which is very short (lifetime $\sim\tau/r$, where τ is the spontaneous emission lifetime for a single atom), and, therefore, has a very broad spectral width $\Delta\omega \sim r\gamma$.

(c) Extension to other cases

So far, we have considered only the case of a gas contained in a volume $\ll\lambda^3$, where λ is the wavelength of the radiation field. What happens when this restriction is removed?

In the first place, different particles can now see different fields, so that we have to go back to (8.3.1) and can no longer make the simplification (8.3.3). In view of the selection rule (8.3.14), it is convenient to expand the field in terms of circularly polarized waves. According to (8.2.3), the operator p^+ associated with absorption (transition $m \to m + 1$) couples with a right-handed circularly polarized wave (this also follows from angular momentum conservation), of time dependence given by

$$E_- = E_x - iE_y = ae^{-i\omega t}, \tag{8.4.22}$$

so that the corresponding field components for a plane wave are

$$E_x(\mathbf{x}, t) = a_\mathbf{k} \cos(\mathbf{k}\cdot\mathbf{x} - \omega t), \qquad E_y(\mathbf{x}, t) = -a_\mathbf{k} \sin(\mathbf{k}\cdot\mathbf{x} - \omega t), \tag{8.4.23}$$

and (8.3.1), together with (8.2.3), yields

$$\mathcal{H}_I = -p \sum_{j=1}^{N} [R^{-(j)} E_+(\mathbf{x}_j, t) + R^{+(j)} E_-(\mathbf{x}_j, t)]$$

$$= -p \sum_\mathbf{k} (a_\mathbf{k}^* \mathcal{R}_\mathbf{k}^- e^{i\omega t} + a_\mathbf{k} \mathcal{R}_\mathbf{k}^+ e^{-i\omega t}), \tag{8.4.24}$$

where

$$\mathcal{R}_\mathbf{k}^\pm = \sum_{j=1}^{N} R^{\pm(j)} \exp(\pm i\mathbf{k}\cdot\mathbf{x}_j) = \mathcal{R}_{\mathbf{k},1} \pm i\mathcal{R}_{\mathbf{k},2}. \tag{8.4.25}$$

It follows from (8.3.2) and (8.4.25) that the operators $\mathcal{R}_{\mathbf{k},1}, \mathcal{R}_{\mathbf{k},2}$ and $\mathcal{R}_3$ [the latter defined by (8.3.4)] obey the angular momentum commutation rules, so that we can define eigenstates $|r_\mathbf{k}, m\rangle$ of $\mathcal{R}_3$ and

$$\mathcal{R}_\mathbf{k}^2 = \mathcal{R}_{\mathbf{k},1}^2 + \mathcal{R}_{\mathbf{k},2}^2 + \mathcal{R}_3^2. \tag{8.4.26}$$

Thus, we can again find superradiant states, but now for radiation in a *given* direction $\mathbf{k}$. What happens for radiation in other directions? It follows from the commutation relations[9] between $\mathscr{R}_{\mathbf{k},i}$ and $\mathscr{R}_{\mathbf{k}',j}$ for $\mathbf{k}' \neq \mathbf{k}$ that the selection rules for the emission or absorption of a photon of wave vector $\mathbf{k}'$ from a system in a state $|r_{\mathbf{k}}, m\rangle$, for $\mathbf{k}' \neq \mathbf{k}$, are

$$\Delta r = \pm 1, 0; \quad \Delta m = \pm 1. \tag{8.4.27}$$

Emission of radiation in directions other than $\mathbf{k}$ tends to produce transitions to states of lower cooperation number, thereby destroying the coherence of atoms in a superradiant state associated with the direction $\mathbf{k}$. This can be compared with an array of classical dipole oscillators phased in such a way that they radiate coherently in a special direction.

One can also interpret the above effect in terms of boson clustering, as mentioned in § 8.1. The emission of a photon in the direction of $\mathbf{k}$ tends to favor the emission of successive photons in the same direction. As an example, if all atoms are initially excited and the first photon is emitted in the direction $\mathbf{k}$, the emission rate for a second photon to be emitted in the same direction is, by (8.4.2), with $r = r_{\mathbf{k}} = N/2$, $m = (N/2) - 1$, given by $I(\mathbf{k}) = 2(N - 1) I_0(\mathbf{k})$, wich is twice the incoherent rate. More generally, the probability of emission in the direction $\mathbf{k}$ contains an enhancement factor equal to the number of photons previously emitted in this direction plus 1. This can be described as an angular correlation between successive emissions of photons.

For a system with dimensions large compared to the wavelength, this enhancement factor should be weighted with another factor representing the directivity of the radiation pattern. The latter factor is strongly dependent on the geometrical shape of the system. Thus, for a spherical container, there is no preferred direction for the emission of the first photon. However, if the container is in the form of an elongated bar, like a cylindrical antenna, the radiation tends to be emitted along directions[10] close to the longitudinal one ("end-fire modes").

8.5 PHOTON ECHOES

How can one excite a system into a superradiant state? One possibility would be to start with a completely inverted population (all active atoms pumped to the excited state), which would correspond to $r = m = N/2$. For a sample with dimensions much smaller than the wavelength, the system would then cascade down by radiation, making transitions to successively lower values of m without changing r [cf. (8.4.1)], and a superradiant state, with m close to zero, would eventually be reached. This assumes that the effects of relaxation mechanisms associated with inhomogeneous broadening can be neglected.

Another possibility would be to start with the system in its ground state, corresponding to $m = -N/2 = -r$, and then to excite it by a radiation pulse sufficiently intense and sufficiently short to raise it to the neighborhood of $m = 0$ without changing r. This second method has actually been realized in practice, overcoming the difficulties due to relaxation, by the observation of *photon echoes*[11]. Similar effects (spin echoes) were known[12] in the case of spins.

A schematic diagram of the experiment is shown in Fig. 8.4. An optically thin ruby crystal was excited by two successive coherent pulses 1 and 2, of width τ in time, separated by a time interval τ_s, obtained by splitting a pulse from a ruby laser and introducing a time delay. For sufficiently high intensities of the incoming pulses, the output contains, in addition to the transmitted pulses 1' and 2', a third pulse 3', the photon echo, which appears a time τ_s after the second transmitted pulse. The pulse widths were of the order of 10 nsec and the interval τ_s between pulses was of the order of 100 nsec.

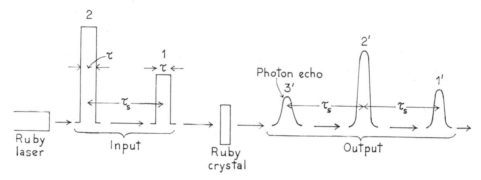

Figure 8. 4 Schematic diagram of the photon echo experiment

In order to explain the formation of the photon echo, let us consider first the case of a sample with dimensions $\ll \lambda$, where λ is the wavelength associated with the laser field. We can then apply the precession equation (8.4.8). Let the electric field be a circularly polarized wave, of time dependence given by (8.4.22),

$$E_x + iE_y = Ee^{i\omega t}. \tag{8.5.1}$$

As is usually done in the theory of magnetic resonance, it is convenient to go over to a new frame of reference, rotating about the z-axis with angular velocity ω; in this frame the electric field (8.5.1) appears "frozen" along a direction that may be taken as the x-axis[13]. The time derivative $\partial/\partial t$ of a vector in the rotating frame is related with the time derivative d/dt in the fixed frame by the well-known relation

$$\frac{d}{dt}\langle \mathscr{P} \rangle = \frac{\partial}{\partial t}\langle \mathscr{P} \rangle + \omega \hat{\mathbf{z}} \times \langle \mathscr{P} \rangle, \tag{8.5.2}$$

so that, taking into account (8.2.14), (8.4.8) becomes

$$\frac{\partial}{\partial t} \langle \mathscr{P} \rangle = \varkappa \langle \mathscr{P} \rangle \times \left(E\hat{\mathbf{x}} + \frac{\Delta\omega}{\varkappa} \hat{\mathbf{z}} \right), \qquad (8.5.3)$$

where

$$\Delta\omega = \omega - \omega_0. \qquad (8.5.4)$$

This discussion assumes that the transition frequency ω_0 is the same for all two-level atoms. Actually, in a ruby crystal, the transition frequency varies slightly from atom to atom, due to variations in local fields; this gives rise to *inhomogeneous broadening*, just like the Doppler effect for atoms in a gas (§ 5.3). It follows that, in (8.3.1), we have to make the substitution

$$\hbar\omega_0 \sum_{j=1}^{N} R_3^{(j)} \to \hbar \sum_{j=1}^{N} \omega_{0j} R_3^{(j)}, \qquad (8.5.5)$$

where ω_{0j} is the transition frequency for the jth atom. By (8.2.16), the pseudo-dipole moment for each atom will still precess around the corresponding pseudoelectric field, but the precession frequency now varies from atom to atom, so that, in the rotating frame, (8.2.16) becomes, for the jth atom,

$$\frac{\partial}{\partial t} \langle \mathscr{P}_j \rangle = \varkappa \langle \mathscr{P}_j \rangle \times \left(E\hat{\mathbf{x}} + \frac{\Delta\omega_j}{\varkappa} \hat{\mathbf{z}} \right), \quad \Delta\omega_j = \omega - \omega_{0j}, \quad (8.5.6)$$

and

$$\langle \mathscr{P} \rangle = \sum_{j=1}^{N} \langle \mathscr{P}_j \rangle. \qquad (8.5.7)$$

The formation of a photon echo is illustrated in Fig. 8.5. Initially, before the first pulse reaches the sample, we can assume that all active atoms are in their ground state, so that all $\langle \mathscr{P}_j \rangle$, as well as their resultant $\langle \mathscr{P} \rangle$, lie along the negative z-axis at $t = 0$ [Fig. 8.5(a)].

Between $t = 0$ and $t = \tau$, we have $E = E_1$ in (8.5.6), corresponding to the electric field amplitude of the first pulse. Thus, each $\langle \mathscr{P}_j \rangle$ precesses about the corresponding pseudo-field $E_1\hat{\mathbf{x}} + \frac{\Delta\omega_j}{\varkappa} \hat{\mathbf{z}}$. We assume that $E_1 \gg |\Delta\omega_j|/\varkappa$, so that all $\langle \mathscr{P}_j \rangle$, as well as their resultant $\langle \mathscr{P} \rangle$, precess about the x-axis by an angle $\varkappa E_1\tau$. The pulse intensity is chosen in such a way that

$$\varkappa E_1\tau = \pi/2; \qquad (8.5.8)$$

such a pulse is called a "$\pi/2$ pulse". It follows that, at $t = \tau$, $\langle \mathscr{P} \rangle$ lies along the y-axis [Fig. 8.5(a)].

Immediately after the passage of the first pulse, each $\langle \mathscr{P}_j \rangle$ starts precessing about the corresponding pseudo-field $[(\Delta\omega_j/\varkappa) \hat{\mathbf{z}}]$. Now the velocity of

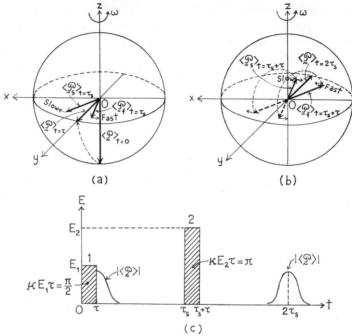

Figure 8.5 Development of a photon echo. (a) Effect of a $\pi/2$ pulse at $t = 0$ on the pseudo-dipole moment, followed by free precession. (b) Effect of a π pulse at $t = \tau_s$, followed by free precession. At $t = 2\tau_s$, the slow and fast components rephase, giving rise to the photon echo. (c) Time development of the pseudo-dipole moment and the electric field of the incident pulses

precession is different for each atom: for a fast atom ($\Delta\omega_f < 0$), the associated $\langle \mathscr{P}_f \rangle$ rotates around the z-axis in the opposite sense from the vector $\langle \mathscr{P}_s \rangle$ associated with a slow atom ($\Delta\omega_s > 0$) [cf. Fig. 8.5(a)]. Thus, the components $\langle \mathscr{P}_j \rangle$ get rapidly out of phase and the magnitude $|\langle \mathscr{P} \rangle|$ of the resultant decreases [Fig. 8.5(c)].

At time $t = \tau_s$, we apply the second pulse, which is made four times as intense as the first one, so that

$$\varkappa E_2 \tau = \pi. \qquad (8.5.9)$$

The effect of this "π pulse" is to reflect each vector $\langle \mathscr{P}_j \rangle$ in the (x, y) plane about the x-axis [Fig. 8.5(b)]. After the passage of the second pulse ($t \geq \tau_s + \tau$), each $\langle \mathscr{P}_j \rangle$ continues to precess around the z-axis with the same angular velocity and in the same sense as it was doing before the second pulse. Therefore, after a time exactly equal to what it had taken them to dephase, i.e., for $t = 2\tau_s$, all components will have rephased again, and we get back the large resultant $\langle \mathscr{P} \rangle$, this time directed along the negative y-axis [Fig. 8.5(b) and (c)].

As we have seen in § 8.4(b), a vector $\langle \mathscr{P} \rangle$ of large magnitude lying in the equatorial plane corresponds to a superradiant state. The spontaneous radiation from this state gives rise to the photon echo, peaked around $t = 2\tau_s$.

A superradiant state is also formed at $t = \tau$ by the first $\pi/2$ pulse [cf. Fig. 8.5(a) and (c)]. However, the radiation from this state has not been observed[14], due to the difficulty of resolving it from the intense overlapping excitation pulse.

The more realistic case in which the dimensions of the sample are large compared with λ can be treated along the lines indicated in § 8.4(c). As was seen there, one can still excite superradiance along the direction of propagation, but the emission rate falls off quite rapidly away from this direction. The radiation pattern therefore tends to become highly directional; it may be identified[1] with the classical radiation pattern of a set of oscillators excited by a plane wave. The result[15] is that the effective volume that becomes superradiant is determined by a kind of "coherence area" of magnitude $\sim \lambda^2$ [cf. (1.3.42) and § 1.2], so that it is of the order of $\lambda^2 l$, where l is the length of the sample.

8.6   SELF-INDUCED TRANSPARENCY

(a) The McCall-Hahn equations

In the experiments on photon echoes, the optical pulses travel through a very thin crystal, so that absorption effects are small. What happens in the other extreme, when an intense short laser pulse travels through a thick resonant medium? Offhand, it might be expected that such pulses would be strongly absorbed, and this, indeed, would be the prediction of the usual linear theory of dispersion. However, it was demonstrated by McCall and Hahn[16], both theoretically and experimentally, that, for sufficiently intense pulses of suitable shape, the medium, due to nonlinear effects, can behave as if it were completely transparent! This remarkable phenomenon, known as *self-induced transparency*, arises when the electric field in the pulse can excite a coherent superposition of the two possible atomic states, taking each atom from the ground state to an excited superposition state and then back again to the ground state. In terms of the "equivalent pendulum" problem described in § 8.4(b), this corresponds to a full pendulum swing by 2π, and to a hyperbolic-secant shape [cf. (8.4.17)], so that the self-transparent pulse is referred to as a "2π h.s." ("2π hyperbolic secant") pulse. The energy absorbed from the leading edge of the pulse to take an atom from the ground state to the excited superposition state is restored to the field, by induced emission, to form the trailing edge (second half of the pulse). The temporary storage of energy in the resonant medium gives rise to a velocity of propagation V for the pulse different from the phase velocity of light $U = c/n$ in the host medium of refractive index n.

In order to analyze the effect, we employ a semiclassical, self-consistent field method similar to that of Lamb's semiclassical theory of the laser[17]. We consider an unbounded host medium of refractive index n, in which are embedded N active two-level atoms per unit volume. As in (8.5.5), we assume that the emission line in the two-level transition is *inhomogeneously broadened*, due to inhomogeneities in crystal fields (or to the Doppler effect in gases). Let $Ng(\gamma)\,d\gamma$ be the number of active atoms per unit volume having a transition frequency between $\omega_0 + \gamma$ and $\omega_0 + \gamma + d\gamma$ (γ plays the role of $\Delta\omega_j$ in (8.5.6), which is now being replaced by a continuous variable). We have

$$\int_{-\infty}^{\infty} g(\gamma)\,d\gamma = 1, \tag{8.6.1}$$

and $g(\gamma)$ plays the role of atomic line shape factor due to inhomogeneous broadening [cf. (5.3.6)]. Usually, $g(\gamma)$ is taken as a symmetric peak centered at $\gamma = 0$, of width

$$\Delta\gamma \sim 1/g(0) \sim 1/T_2^*, \tag{8.6.2}$$

where T_2^* is the lifetime associated with inhomogeneous broadening (the time taken by atomic dipoles to dephase in § 8.5).

We want to consider the propagation of a circularly polarized pulse [cf. (8.4.23)],

$$\mathbf{E}(z, t) = \mathscr{E}(z, t)\,\{\cos[kz - \omega t + \varphi(z)]\,\hat{\mathbf{x}} - \sin[kz - \omega t + \varphi(z)]\,\hat{\mathbf{y}}\}, \tag{8.6.3}$$

where

$$k = \omega/U = n\omega/c, \tag{8.6.4}$$

and the "amplitude" $\mathscr{E}(z, t)$ and the "phase" $\varphi(z)$ are *slowly-varying functions* [cf. (6.2.13)], so that

$$|\partial\mathscr{E}/\partial z| \ll k\,|\mathscr{E}|, \quad |\partial\mathscr{E}/\partial t| \ll \omega\,|\mathscr{E}|, \quad |\partial\varphi/\partial z| \ll k. \tag{8.6.5}$$

We ignore a possible dependence of φ on t, thus neglecting frequency modulation and pulling effects. For a short pulse, the frequency ω is not well-defined, and, without loss of generality, we can set it equal to the central resonance frequency,

$$\omega = \omega_0. \tag{8.6.6}$$

Henceforth we shall drop the index 0 and just employ the notation ω for the common frequency.

In order to extend (8.5.6), we make the transformation from the medium rest frame (x, y, z) to the rotating frame (x', y', z) in such a way as to eliminate the phase dependence of the wave:

$$x + iy = (x' + iy') \exp [i(\omega t - kz - \varphi)], \qquad (8.6.7)$$

so that we have a different rotating frame for each value of z. It then follows from (8.6.3) that

$$E_x' + iE_y' = (E_x + iE_y) \exp [i(kz - \omega t + \varphi)] = \mathscr{E}(z, t), \qquad (8.6.8)$$

so that $E_x' = \mathscr{E}$, $E_y' = 0$, and we may therefore employ (8.5.6). For an atom with a transition frequency $\omega + \gamma$, we get

$$\frac{\partial}{\partial t} \langle \mathscr{P}' \rangle = \varkappa \langle \mathscr{P}' \rangle \times \left[\mathscr{E}(z, t)\hat{\mathbf{x}}' + \frac{\gamma}{\varkappa} \hat{\mathbf{z}} \right]. \qquad (8.6.9)$$

By (8.6.7), we have, in the medium rest frame,

$$\mathscr{P}_+ = \mathscr{P}_x + i\mathscr{P}_y = \mathscr{P}_+' \exp [i(\omega t - kz - \varphi)]. \qquad (8.6.10)$$

As there are $Ng(\gamma)\, d\gamma$ atoms per unit volume with transition frequencies between $\omega + \gamma$ and $\omega + \gamma + d\gamma$, the macroscopic polarization (cf. § 6.4) is given by

$$P_+(z, t) = P_x(z, t) + iP_y(z, t)$$
$$= \int_{-\infty}^{\infty} g(\gamma) [u(\gamma, z, t) + iv(\gamma, z, t)] \exp [i(\omega t - kz - \varphi)]\, d\gamma, \qquad (8.6.11)$$

where u and v, the dispersive and absorptive components of the polarization, are given by

$$u(\gamma, z, t) + iv(\gamma, z, t) = N[\langle \mathscr{P}_x'(\gamma, z, t) \rangle + i\langle \mathscr{P}_y'(\gamma, z, t) \rangle]. \qquad (8.6.12)$$

By (8.2.13) and (8.3.1),

$$w(\gamma, z, t) = N\langle \mathscr{P}_z'(\gamma, z, t) \rangle = (N\varkappa/\omega) \langle \mathscr{H}_0 \rangle = \varkappa W/\omega, \qquad (8.6.13)$$

where $W = N\langle \mathscr{H}_0 \rangle$ is the spectral energy density of excitation of the active medium.

From (8.6.9), (8.6.12) and (8.6.13), we get the equations of motion for the macroscopic pseudo-polarization $u\hat{\mathbf{x}}' + v\hat{\mathbf{y}}' + w\hat{\mathbf{z}}$, which will play the role of "pendulum vector",

$$\frac{\partial u}{\partial t} = \gamma v, \qquad (8.6.14)$$

$$\frac{\partial v}{\partial t} = -\gamma u + \varkappa w\mathscr{E}, \qquad (8.6.15)$$

$$\frac{\partial w}{\partial t} = -\varkappa v\mathscr{E}. \qquad (8.6.16)$$

It follows from these equations that $u^2 + v^2 + w^2$ is conserved. This is just the expression (6.3.44) of probability conservation. In fact, by (8.2.11) and (8.2.13),

$$u^2 + v^2 + w^2 = N^2 p^2. \qquad (8.6.17)$$

All line-broadening effects other than inhomogeneous broadening have been neglected in the above discussion. A phenomenological description of other effects may be obtained by adding Bloch-type[18] damping terms,

$$\left(\frac{\partial u}{\partial t}\right)_{\text{damping}} = -\frac{u}{T_2'}, \qquad \left(\frac{\partial v}{\partial t}\right)_{\text{damping}} = -\frac{v}{T_2'},$$

$$\left(\frac{\partial w}{\partial t}\right)_{\text{damping}} = -\frac{(w - w_0)}{T_1}, \qquad (8.6.18)$$

where T_1 is the energy damping lifetime associated with relaxation to the ground state value (corresponding to $W_0 = -N\hbar\omega/2$)

$$w_0 = -Np, \qquad (8.6.19)$$

and T_2' includes all dipole-moment damping effects other than inhomogeneous broadening (in particular, $1/T_2'$ includes a contribution from $1/T_1$). The total linewidth $1/T_2$ is approximately given by

$$1/T_2 \sim 1/T_2' + 1/T_2^*. \qquad (8.6.20)$$

We will consider the propagation of pulses of temporal width τ such that

$$\omega^{-1} \ll \tau \ll T_2', \qquad (8.6.21)$$

so that we can set

$$T_1 \to \infty, \qquad T_2' \to \infty. \qquad (8.6.22)$$

and neglect the damping terms (8.6.18), as has been done in (8.6.14)–(8.6.16). On the other hand, T_2^* can be either smaller or larger than τ.

The macroscopic polarization acts like a source for the field. By Maxwell's equations, we have, in the medium rest frame,

$$\frac{\partial^2 E_+}{\partial z^2} - \frac{1}{U^2} \frac{\partial^2 E_+}{\partial t^2} = \frac{4\pi}{c^2} \frac{\partial^2 P_+}{\partial t^2}, \qquad (8.6.23)$$

which corresponds to (6.2.7) with $\sigma = 0$ (we have switched over to Gaussian units). Here, P_+ is given by (8.6.11), and, according to (8.6.8),

$$E_+ = E_x + iE_y = \mathscr{E}(z, t) \exp\left[i(\omega t - kz - \varphi)\right]. \qquad (8.6.24)$$

Substituting (8.6.11) and (8.6.24) in (8.6.23), neglecting terms that are small by the slowly-varying amplitude and phase approximation (8.6.5), and identifying real and imaginary parts, we find

$$\frac{\partial \mathscr{E}}{\partial z} + \frac{1}{U}\frac{\partial \mathscr{E}}{\partial t} = 2\pi\frac{\omega}{nc}\int_{-\infty}^{\infty} v(\gamma, z, t)\,g(\gamma)\,dy, \tag{8.6.25}$$

$$\frac{\partial \varphi}{\partial z}\mathscr{E} = 2\pi\frac{\omega}{nc}\int_{-\infty}^{\infty} u(\gamma, z, t)\,g(\gamma)\,dy. \tag{8.6.26}$$

These results, together with the "Ansatz" (8.6.3), imply the absence of backscattered radiation: we will have only a forward-traveling pulse. According to Crisp[19], the neglect of backscattering is a very good approximation for the self-transparent solutions.

The system of nonlinear equations (8.6.14)–(8.6.16), (8.6.25) and (8.6.26) will be referred to as the McCall-Hahn equations. We now demonstrate the existence of self transparent solutions of these equations.

(b) The hyperbolic-secant solution

A self-transparent pulse is a solution of the McCall-Hahn equations such that the pulse envelope $\mathscr{E}(z, t)$ travels without change of shape with velocity V. Therefore, let us look for solutions of the form

$$\mathscr{E} = \mathscr{E}(s), \quad u = u(\gamma, s), \quad v = v(\gamma, s), \quad w = w(\gamma, s), \tag{8.6.27}$$

where

$$s = t - z/V. \tag{8.6.28}$$

We indicate time derivatives (which are the same as derivatives with respect to s) by dots. Furthermore, it seems reasonable to assume that all two-level atoms respond in the same manner (i.e., with the same time dependence) to the pulse, but with different amplitudes, depending on their detuning γ from the resonance frequency ω, so that we look for a "separated" solution in which the polarization is of the form

$$v(\gamma, s) = v(0, s)\,f(\gamma), \tag{8.6.29}$$

where we have normalized the frequency-dependent factor by

$$f(0) = 1. \tag{8.6.30}$$

We will see that the other polarization component $u(\gamma, s)$ is also of the same "factored" form.

Substituting (8.6.27) and (8.6.29) in (8.6.25), we find

$$a\mathscr{E} = -v(0, s), \tag{8.6.31}$$

where

$$a = \frac{nc[(1/V) - (1/U)]}{2\pi\omega F},$$ (8.6.32)

$$F = \int_{-\infty}^{\infty} f(\gamma) \, g(\gamma) \, d\gamma.$$ (8.6.33)

Combining (8.6.29), (8.6.31), (8.6.14) and (8.6.16), we get

$$\dot{u} = -a\gamma f \dot{\mathscr{E}},$$ (8.6.34)

$$\dot{w} = \varkappa a f \mathscr{E} \dot{\mathscr{E}}.$$ (8.6.35)

In order to integrate these equations, we take as initial conditions the values of the quantitites in the remote past $(t \to -\infty)$; we assume that the active medium was originally unpolarized, with no field present:

$$\lim_{s \to -\infty} u(\gamma, s) = \lim_{s \to -\infty} v(\gamma, s) = \lim_{s \to -\infty} \mathscr{E}(s) = 0.$$ (8.6.36)

It then follows from (8.6.17) that [cf. (8.6.19)]

$$\lim_{s \to -\infty} w(\gamma, s) = \pm w_0 = \mp Np.$$ (8.6.37)

The solutions with initial value w_0 correspond to the active medium originally in the ground state, and will be referred to as "attenuator" solutions; those with initial value $-w_0$ correspond to the active medium originally in the completely inverted state (all atoms excited), and will be referred to as "amplifier" solutions. The stability of the completely inverted state is a consequence of the neglect of damping effects [cf. (8.6.22)], and it is clearly not a realistic assumption for times comparable with the energy damping lifetime T_1 [cf. (8.6.18)].

Given any solution of the above form, we obtain another solution by making the substitutions[20]

$$\mathscr{E} \to \mathscr{E}, \quad u \to -u, \quad v \to -v, \quad w \to -w, \quad a \to -a.$$ (8.6.38)

Thus, it is sufficient to confine our attention to attenuator solutions, taking the upper signs in (8.6.37).

Integrating (8.6.34) and (8.6.35), under these conditions, we find two additional "conservation laws",

$$u(\gamma, s) = -a\gamma f(\gamma) \, \mathscr{E}(s),$$ (8.6.39)

$$w(\gamma, s) - w_0 = \tfrac{1}{2}\varkappa a f(\gamma) \, \mathscr{E}^2(s).$$ (8.6.40)

Note that $u(\gamma, s)$ appears indeed in factored form.

RESONANCE CASE Let us consider first the atoms at resonance, for which $\gamma = 0$. It then follows from (8.6.39) that

$$u(0, s) = 0, \tag{8.6.41}$$

so that the "pendulum vector" remains in the (v, w) plane. By (8.6.17), we may set

$$v(0, s) = Np \sin (\pi - \varphi) = Np \sin \varphi,$$
$$w(0, s) = Np \cos (\pi - \varphi) = -Np \cos \varphi, \tag{8.6.42}$$

where the angle φ measures the deviation from the vertical stable equilibrium position $\varphi_0 = 0$. We now get from (8.6.30) and (8.6.40)

$$a\mathscr{E}^2(s) = \frac{2}{\varkappa} Np(1 - \cos \varphi) = \frac{4Np}{\varkappa} \sin^2 \frac{\varphi}{2}, \tag{8.6.43}$$

from which we conclude that u is non-negative.

Taking the square root of both sides of (8.6.43) and substituting it, together with (8.6.42), in (8.6.31) we get[21]

$$\mathscr{E}(s) = -2\sqrt{(Np/\varkappa a)} \sin \frac{\varphi}{2}, \tag{8.6.44}$$

$$\frac{\dot\varphi}{2} = \frac{1}{\tau} \sin \frac{\varphi}{2}, \tag{8.6.45}$$

where

$$\tau = \sqrt{(a/Np\varkappa)}. \tag{8.6.46}$$

We now recognize (8.6.45) as the pendulum equation (8.4.12), with φ given by (8.4.13). The solution is therefore given by (8.4.17),

$$\varphi(s) = 4 \tan^{-1}(e^{s/\tau}), \quad \sin \frac{\varphi}{2} = \operatorname{sech}(s/\tau). \tag{8.6.47}$$

From (8.6.44) to (8.6.47), it follows that

$$\mathscr{E}(s) = -\frac{\dot\varphi}{\varkappa} = -\frac{2}{\varkappa\tau} \sin \frac{\varphi}{2} = -\frac{2}{\varkappa\tau} \operatorname{sech}(s/\tau), \tag{8.6.48}$$

and, integrating both sides with respects to t from $-\infty$ to ∞,

$$-\varkappa \int_{-\infty}^{\infty} \mathscr{E}\left(t - \frac{z}{V}\right) dt = \varphi(\infty) - \varphi(-\infty) = 2\pi. \tag{8.6.49}$$

Thus, the magnitude[22] of the area under the pulse, multiplied by $\varkappa$, is equal to 2π, the total "pendulum swing angle" of the pseudo-polarization for resonant atoms. This is the reason for the name "2π hyperbolic-secant pulse" given to the self-transparent solution (8.6.48).

OFF-RESONANCE CASE   For $\gamma \neq 0$, $\mathscr{E}(s)$, which does not depend on γ, must still be given by (8.6.48), and (8.6.42), together with the "Ansatz" (8.6.29), implies

$$v(\gamma, s) = Npf(\gamma) \sin \varphi, \tag{8.6.50}$$

whereas (8.6.39) and (8.6.40) become [(taking into account (8.6.19)]

$$u(\gamma, s) = \frac{2a\gamma}{\varkappa\tau} f(\gamma) \sin \frac{\varphi}{2}, \tag{8.6.51}$$

$$w(\gamma, s) = -Np + \frac{2a}{\varkappa\tau^2} f(\gamma) \sin^2 \frac{\varphi}{2}. \tag{8.6.52}$$

We must still satisfy condition (8.6.17). Substituting the above expressions for u, v, w, and taking into account (8.6.46), we find that (8.6.17) is satisfied if and only if

$$f(\gamma) = \frac{1}{1 + \tau^2\gamma^2}. \tag{8.6.53}$$

Thus, the amplitude of the off-resonance absorptive part of the polarization, as well as the off-resonance atomic excitation energy $W - W_0 = \dfrac{\omega}{\varkappa} \times$

$\times (w + Np)$, falls off with the detuning γ according to a Lorentzian law, with halfwidth $\delta\gamma$ equal to the reciprocal pulse width τ^{-1}.

Substituting (8.6.53) in the above expressions, we find the explicit self-transparent solution of the McCall-Hahn equations,

$$u(\gamma, z, t) = \frac{2Np\tau\gamma}{1 + \tau^2\gamma^2} \sin \frac{\varphi}{2}, \tag{8.6.54}$$

$$v(\gamma, z, t) = \frac{Np}{1 + \tau^2\gamma^2} \sin \varphi, \tag{8.6.55}$$

$$w(\gamma, z, t) + Np = \frac{\varkappa}{\omega} (W - W_0) = \frac{2Np}{1 + \tau^2\gamma^2} \sin^2 \frac{\varphi}{2}, \tag{8.6.56}$$

$$\mathscr{E}(z, t) = - \frac{2}{\varkappa\tau} \sin \frac{\varphi}{2}, \tag{8.6.57}$$

where

$$\sin \frac{\varphi}{2} = \operatorname{sech} \left[\frac{1}{\tau} \left(t - \frac{z}{V} \right) \right]. \tag{8.6.58}$$

We may have 2π h.s. pulses of any width τ, but the pulse velocity V is related to the width by (8.6.46), (8.6.32), (8.6.33),

$$\frac{1}{V} - \frac{1}{U} = \frac{2\pi\omega}{nc} aF = \frac{2\pi\omega}{nc} Np\varkappa\tau^2 \int_{-\infty}^{\infty} \frac{g(\gamma)}{1 + \tau^2\gamma^2} d\gamma. \tag{8.6.59}$$

There still remains the equation (8.6.26) for the phase. Substituting (8.6.54) and (8.6.57), we find

$$\frac{\partial \varphi}{\partial z} = - \frac{2\pi\omega}{nc} N p\varkappa\tau^2 \int_{-\infty}^{\infty} \frac{\gamma g(\gamma)}{1 + \tau^2\gamma^2} \, d\gamma = k', \qquad (8.6.60)$$

so that

$$\varphi(z) = k'z + \varphi_0. \qquad (8.6.61)$$

According to (8.6.3), this has the effect of changing the wave number from k to $k + k'$, which can be interpreted as a change in refractive index due to the presence of the active medium. With the usual assumption of a symmetric line shape function $g(\gamma)$, we have $k' = 0$, and it is then consistent to take $\varphi = 0$. The results (8.6.54)–(8.6.56) show the remarkable feature that, regardless of the detuning γ and the line shape $g(\gamma)$, all two-level atoms start from the ground-state values $u = 0$, $v = 0$, $W = W_0$, are temporarily excited by the pulse into a coherent superposition of the upper and lower states and then go back to the ground-state values. The curves described by the tip of the pseudo-polarization vector on the sphere of radius Np for some values of $\gamma\tau$ are shown in Fig. 8.6.

We see from (8.6.56) that, during the first half of the pulse ($-\infty < t \leq z/V$), the active medium draws energy from the field; during the second half, the same energy is returned to the field. The temporary storage of en

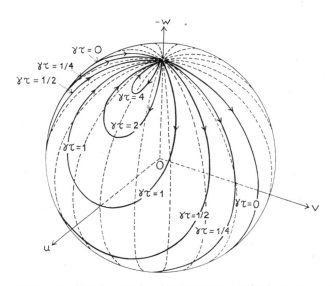

Figure 8.6 Curves described by the tip of the pseudo-polarization vector on the sphere of radius Np as a function of $\gamma\tau$ [after S. L. McCall and E. L. Hahn, *Phys. Rev.* **183**, 457 (1969)]

13*

ergy in the active medium delays the pulse, giving rise to a pulse velocity $V < U$. According to (8.6.59), the longer the pulse, the more slowly it will travel; very sharp pulses ($\tau \to 0$) tend to travel at the phase velocity U of light in the host medium.

The result (8.6.59) has a very simple physical interpretation[23] in terms of energy balance. The energy flow takes place *only* through the electromagnetic wave, with velocity U. Thus, the total energy per unit area per unit time flowing through a section of the medium is given by the magnitude of Poynting's vector,

$$S = U W_{em} = U \frac{n^2}{4\pi} \mathscr{E}^2, \qquad (8.6.62)$$

where W_{em} is the electromagnetic energy density, and we have used the fact that, in the slowly-varying envelope and phase approximation, the relation between $\mathbf{E}$ and $\mathbf{B}$ in the pulse is the same as in the host medium. On the other hand, since the pulse envelope travels with velocity V, this must also be the *total* energy contained in a cylinder of unit cross-section and length V, i.e.,

$$S = V(W_{em} + \Delta W), \qquad (8.6.63)$$

where ΔW is the total *excitation* energy density stored in the two-level atoms, which is given by

$$\Delta W = \int_{-\infty}^{\infty} g(\gamma) \, (W - W_0) \, d\gamma. \qquad (8.6.64)$$

Equating the two expressions for S, we get

$$\frac{1}{V} - \frac{1^{\cdot}}{U} = \frac{1}{U} \frac{\Delta W}{W_{em}}. \qquad (8.6.65)$$

Taking into account (8.6.62), (8.6.64), (8.6.13), and (8.6.40), this becomes

$$\frac{1}{V} - \frac{1}{U} = \frac{(\omega/2) \, a \mathscr{E}^2 F}{(n^2/4\pi) \, U \mathscr{E}^2} = \frac{2\pi\omega}{nc} \, aF,$$

which is the same as (8.6.59). Thus, the difference (8.6.65) between the inverse velocities is proportional to the ratio of the excitation energy stored in the active medium to the electromagnetic energy. We see also that the "conservation law" (8.6.40) is an expression of the energy balance.

(c) The area theorem

So far we have not discussed what happens to pulses that are not of the 2π h.s. type. Some insight into this may be obtained from an important result due to McCall and Hahn, the *area theorem*. Let us define

$$\varphi(z) = -\varkappa \int_{-\infty}^{\infty} \mathscr{E}(z, t) \, dt. \qquad (8.6.66)$$

This quantity is proportional to the total area under the pulse envelope at a given z, for a pulse of arbitrary shape. For resonant atoms ($\gamma = 0$), $u(\gamma, z, t)$ vanishes, according to (8.6.14), so that (8.6.42) is still valid for an arbitrary pulse, with $\varphi = \varphi(z, t)$:

$$v(0, z, t) = Np \sin \varphi(z, t), \qquad w(0, z, t) = -Np \cos \varphi(z, t), \qquad (8.6.67)$$

where, by (8.6.15) and (8.6.16),

$$\frac{\partial}{\partial t} \varphi(z, t) = -\varkappa \mathscr{E}(z, t), \qquad (8.6.68)$$

so that

$$\varphi(z, \infty) - \varphi(z, -\infty) = -\varkappa \int_{-\infty}^{\infty} \mathscr{E}(z, t)\, dt. \qquad (8.6.69)$$

Comparing (8.6.69) with (8.6.66), we see that $\varphi(z)$ also represents the *total tipping angle* of the pseudo-polarization vector for *resonant atoms* at z.

Multiplying both sides of (8.6.25) by $-\varkappa$, making use of (8.6.14) and integrating over t from $-\infty$ to ∞, we get, by (8.6.66),

$$\frac{d\varphi}{dz} = -\frac{2\pi\omega\varkappa}{nc} \lim_{T \to \infty} \int_{-\infty}^{T} g(\gamma)\,u(\gamma, z, T)\frac{d\gamma}{\gamma}, \qquad (8.6.70)$$

where we have assumed that the pulse dies out at both ends and there is no initial polarization,

$$\mathscr{E}(z, \infty) = \mathscr{E}(z, -\infty) = u(\gamma, z, -\infty) = 0. \qquad (8.6.71)$$

Let $T = T_0 + t'$, where T_0 is chosen large enough that we may already set (with negligible error)

$$\mathscr{E}(z, T_0 + t') = 0 \quad \text{for} \quad t' \geq 0. \qquad (8.6.72)$$

This does *not* imply $u = v = 0$ for $t' \geq 0$: it implies only that different frequency components interfere destructively to make the net polarization (8.6.11) vanish (cf. the discussion of dephasing in § 8.5).

For $t' \geq 0$, by (8.6.72), (8.6.14) and (8.6.15), we have free precession around the z-axis, so that

$$u(\gamma, z, T_0 + t') = u(\gamma, z, T_0) \cos (\gamma t') + v(\gamma, z, T_0) \sin (\gamma t'). \qquad (8.6.73)$$

Substituting in (8.6.70), and making use of the well-known results[24]

$$\lim_{t' \to \infty} \left[\frac{\sin (\gamma t')}{\gamma} \right] = \pi \delta(\gamma), \quad \lim_{t' \to \infty} \left[\frac{1 - \cos (\gamma t')}{\gamma} \right] = \frac{\mathscr{P}}{\gamma}, \qquad (8.6.74)$$

where $\mathcal{P}$ denotes the Cauchy principal value, we get

$$\frac{d\varphi}{dz} = -\frac{2\pi^2\omega\varkappa}{nc} g(0)\, v(0, z, T_0)$$

$$-\frac{2\pi\omega\varkappa}{nc}\int_{-\infty}^{\infty}\left(\frac{1}{\gamma} - \frac{\mathcal{P}}{\gamma}\right) u(\gamma, z, T_0)\, g(\gamma)\, d\gamma. \qquad (8.6.75)$$

The integral vanishes identically, because $u = O(\gamma)$ for $\gamma \to 0$ [cf. (8.6.14)], so that the integrand is regular. On the other hand, by (8.6.67), (8.6.69) and (8.6.72),

$$v(0, z, T_0) = Np \sin \varphi(z, T_0) = Np \sin \varphi(z), \qquad (8.6.76)$$

because by assumption $\varphi(z, -\infty) = 0$. Substituting in (8.6.75), we finally get the *area theorem*,

$$\frac{d\varphi}{dz} = -\frac{\alpha}{2} \sin \varphi(z), \qquad (8.6.77)$$

where [cf. (8.2.8)]

$$\alpha = 8\pi^2 \frac{\omega Np^2}{nc\hbar} g(0). \qquad (8.6.78)$$

For small pulse intensities, i.e., in the linear regime, we have

$$\varphi \ll 1, \quad \sin \varphi \approx \varphi, \qquad (8.6.79)$$

and (8.6.77) yields

$$\varphi(z) \approx \varphi(0)\, e^{-\alpha z/2}, \quad \mathscr{E}^2(z) \approx \mathscr{E}^2(0)\, e^{-\alpha z}, \qquad (8.6.80)$$

which is the well-known Beer's law for extinction. We see that the parameter α in (8.6.78) represents the extinction coefficient of the medium at the resonant frequency ω, according to the linear theory of dispersion.

The general solution of (8.6.77) is obtained just as in (8.4.12), (8.4.17),

$$\tan (\varphi/2) = \tan (\varphi_0/2) \exp (-\tfrac{1}{2}\alpha z) \qquad (8.6.81)$$

where $\varphi_0 = \varphi(0)$. The branch solutions of (8.6.81) are plotted in Fig. 8.7(a). Note that the sign of $\tan (\varphi/2)$ is determined by that of $\tan (\varphi_0/2)$, so that φ remains between $n\pi$ and $(n + 1)\pi$ for all z. As $z \to \infty$, we see that $\varphi \to 2n\pi$, so that only pulses with an area that is a multiple of 2π *can* be stable.

What actually happens to a pulse with given initial shape and area has been determined, in a number of examples, by computer calculations. The behavior of initially Gaussian pulses with initial areas of $0.9\,\pi$ and $1.1\,\pi$, respectively, is shown in Fig. 8.7. We see that pulses with initial areas smaller than π evolve towards zero area, whereas pulses of initial area greater than π tend to approach the stable 2π h.s. pulse solution; in this process, they lose

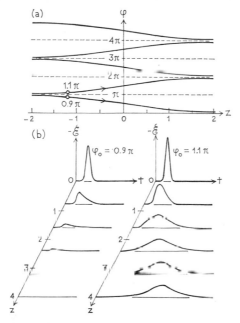

Figure 8.7 (a) Branch solutions of (8.6.81). For $\alpha > 0$, φ evolves in the direction of increasing z towards the nearest even multiple of π. (b) Computer plots of the evolution of initially Gaussian pulses with $\varphi_0 = 0.9\pi$ and $\varphi_0 = 1.1\pi$ as a function of time and of z. The distance z is measured in absorption length units of $\pi\alpha^{-1}$ [after S. L. McCall and E. L. Hahn, *Phys. Rev.* **183**, 457 (1969)]

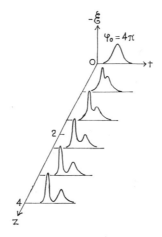

Figure 8.8 Computer plot of the evolution of a pulse with initial area $\varphi_0 = 4\pi$. The pulse splits into two separate 2π pulses, whose widths and separation depend on the initial pulse shape. The distance z is measured in absorption length units of $\pi\alpha^{-1}$ [after S. L. McCall and E. L. Hahn, *Phys. Rev.* **183**, 457 (1969)]

some energy and are reshaped, over a few absorption lengths α^{-1} (remember that, for a pulse of width τ, the energy is roughly $\sim \mathscr{E}^2\tau$, whereas the area is $\sim \mathscr{E}\tau$).

What happens to $2n\pi$ pulses, $n > 1$? Computer calculations indicate that such pulses tend to break up into n separate 2π pulses, with various widths and delay times. An example for $\varphi_0 = 4\pi$ is shown in Fig. 8.8.

Several effects have been disregarded in the above analysis, such as the nonuniform intensity profile of the incident laser pulse, which introduces deviations from the ideal plane-wave model; self-focusing effects; the effects of transverse modes, finite relaxation times and Doppler shifts. For a discussion of the effects due to these additional complications, the reader is referred to the original McCall-Hahn papers.

(d) Discussion

(i) EXPERIMENTS The first experimental observations of self-induced transparency were made by McCall and Hahn, with pulses from a ruby laser traveling through a liquid-helium cooled ruby rod. The pulse widths were of the order of a few n sec; T_2^* was $\sim 10^{-10}$ sec. Subsequently, the effect was observed[25] in a gaseous medium, with pulses from a CO_2 laser propagating through gaseous SF_6, for which there exists a transition resonant with the 10.6 μ CO_2 laser radiation. One of the important characteristics of self-induced transparency, which allows one to distinguish it from other effects, such as the saturation of absorption at high intensities (which would also lead to departures from Beer's law), is the time delay of the output pulse, and its dependence on the pulse width (or intensity, for a given area).

If the inhomogeneous linewidth $\Delta\gamma \sim 1/T_2^*$ is much larger than the width $\delta\gamma \sim \tau^{-1}$ associated with the pulse, the result (8.6.59) becomes [cf. (8.6.78)]

$$\frac{1}{V} - \frac{1}{U} \approx \frac{2\pi^2\omega}{nc} Np\varkappa g(0)\,\tau = \frac{\alpha\tau}{2}, \qquad (8.6.82)$$

If $\alpha\tau \gg 1/U$, this yields $V \approx 2(\alpha\tau)^{-1}$; under these conditions, most of the pulse energy is stored in the resonant medium, and the pulse is retarded by about one pulse width per absorption length α^{-1}.

More recently Gibbs and Slusher[26] have observed self-induced transparency in a gas of Rb atoms, using pulses from a Hg laser. In this case, $T_2^* = 0.8$ nsec, $T_2 \sim 28$ nsec, and the pulse width was 6.5 nsec. For some inputs, the output peak intensity exceeded the input peak intensity. The maximum observed delay was ~ 15–20 nsec, corresponding to a reduction in the velocity of propagation to $\approx c/5{,}000$, and a corresponding pulse contraction in space from ~ 2 m to less than 1 mm! The breakup of 4π and 6π pulses into two and three 2π pulses, respectively, was also observed. The agreement between computed and observed pulse shapes was found to be quite good.

Self-induced transparency has also been observed in acoustic paramagnetic resonance[27]. It provides a new experimental technique for studying short relaxation times.

(ii) AMPLIFYING MEDIUM   Our discussion so far has been restricted to an attenuator, corresponding to the assumption that $\varphi(z, -\infty) = 0$ in (8.6.69). We can treat the case of a completely inverted amplifying medium, in which all active atoms are initially in the excited state ($\varphi(z, -\infty) = \pi$) by substituting φ by $\pi - \varphi$ in the above discussion. According to (8.6.77), this is equivalent to replacing α by $-\alpha$ or z by $-z$.

By reading Fig. 8.7 with φ evolving in the $-z$ direction, we see that, in an amplifying medium, pulses with initial areas between 0 and 2π will tend to evolve into a π pulse. What happens can be understood by reference to Fig. 8.7(b). To begin with, all the atoms are excited. If we now start with a very weak pulse, it follows from (8.6.80) with $\alpha \to -\alpha$ (or $z \to -z$) that such a pulse will be amplified, its area increasing at first exponentially, by drawing energy from the excited atoms. As the pulse area begins to approach π, the pulse is taking resonant atoms from the excited state to the ground state, dumping all their energy into the wavefront. As its energy increases, the pulse must narrow, to keep the area constant[28], thus tending to build up something similar to a shock wave. This effect was first predicted by Dicke[29].

As has already been mentioned, the stability of the completely inverted state in the above treatment follows from the neglect of energy damping effects (including those of spontaneous emission), and it is an unrealistic assumption for times comparable with the lifetime T_1. The losses that occur in a real medium would limit the growth of power in a π pulse. These effects and others have been taken into account in more detailed treatments[30].

(iii) CAUSALITY   According to (8.6.38), a 2π h.s. pulse can also propagate in an amplifying medium; by (8.6.59), its velocity V would then be greater than the phase velocity of light U in the host medium, by an amount that would increase with the pulse width τ. Actually, as can be seen in Fig. 8.7, such a solution would correspond to unstable equilibrium, and it is no longer present if one adds a perturbation (such as the neglected damping effects). However, peak velocities greater than U are also found in other pulse solutions in an amplifying medium.

Since such velocities can exceed c, the question arises as to whether any violation of causality may be involved. Due to the essentially nonlinear character of these phenomena, the usual causality criteria for linear dispersive media[31] cannot be applied.

It is easy to see that the velocity of the peak of the pulse envelope is a purely geometric concept, which, even for a 2π h.s. pulse (which travels without change of shape), is quite different from the velocity of enregy propagation. The point is that, in the case of an amplifying medium, the weak leading edge of a 2π h.s. pulse (which has a tail extending to infinity) triggers the release of the energy stored in the excited atoms, which then gives rise to the peak. Thus, in contrast with the case of an attenuator, the energy is already stored in the medium prior to the arrival of the bulk of the pulse; it is not "the same energy" that travels along with the pulse envelope. This argument suggests that no violation of causality need be involved[32].

It can be rigorously shown[33] that all possible solutions of the McCall-Hahn equations [with or without the damping terms (8.6.18)] satisfy the causality condition, in the sense that influence is not propagated faster than c. The domain of dependence of the solution at a given spacetime point is contained within the backward light cone with vertex at that point.

(iv) MULTIPLE PULSE SOLUTIONS   In addition to the 2π h.s. pulse, the McCall-Hahn equations admit other solutions[34] of the form (8.6.27). However, such solutions do not satisfy conditions (8.6.36): instead of a single pulse that dies out as $t \to \pm\infty$, the field is an infinite periodic wave train, coresponding to the periodic swings or oscillations of the "equivalent pendulum" discussed in § 8.4(b). Such solutions, like infinite plane waves, would have infinite energy.

Analytical solutions describing the splitting of $2n\pi$ pulses (cf. Fig. 8.8) have been obtained by Lamb[35] in the limiting case of an infinitely narrow atomic line [cf. (8.6.1)],

$$g(\gamma) = \delta(\gamma). \tag{8.6.83}$$

In this limit, as in (8.6.41)–(8.6.42), we can set

$$u = 0; \quad v = Np \sin \varphi(z, t); \quad w = Np \cos \varphi(z, t), \tag{8.6.84}$$

and [cf. (8.6.48)]

$$\mathcal{E}(z, t) = (1/\varkappa) \, \partial\varphi/\partial t, \tag{8.6.85}$$

whereas (8.6.25) takes the form

$$\frac{\partial \mathcal{E}}{\partial z} + \frac{1}{U} \frac{\partial \mathcal{E}}{\partial t} = 2\pi \frac{\omega}{nc} Np \sin \varphi. \tag{8.6.86}$$

Substituting (8.6.85) in (8.6.86), and making the change of variables

$$\zeta = z, \quad \tau = t - z/U, \tag{8.6.87}$$

we find

$$\frac{\partial^2 \varphi}{\partial \zeta \partial \tau} = \beta \sin \varphi(\zeta, \tau); \quad \beta = \frac{4\pi\omega Np^2}{nc\hbar}. \tag{8.6.88}$$

The general solution of this nonlinear partial differential equation is unknown (the linear version, with sin $\varphi \approx \varphi$, is readily recognized as a form of the two-dimensional Klein-Gordon equation). However, it is known that, given a particular solution, one may generate other solutions by a transformation known as a Baecklund transformation. By using this device and combining 2π h.s. solutions, Lamb has shown how to generate $2n\pi$ pulse solutions. As has already been mentioned, such pulses split into n separate 2π pulses for sufficiently large times.

It is interesting to note that a similar phenomenon has been found to occur in solutions of another nonlinear partial differential equation, the Korteweg-de Vries equation[36], which also describes propagation in nonlinear dispersive media. For large times, the solution of this equation also tend to break up into a number of isolated pulses known as "solitons", or solitary waves. Such pulses have a well-defined shape and show a remarkable degree of stability, preserving their identity even after nonlinear interactions. We can think of the 2π h.s. pulses as a kind of solitons associated with the McCall-Hahn equations.

8.7   CONCLUDING REMARKS

It may be useful to conclude by summarizing some of the most valuable new contributions to theoretical physics that we feel have been made by the recent developments in quantum optics discussed in these lectures.

(i) COHERENCE   We have learned how to give a quantum-mechanical description of coherence. The coherence functions introduced for this purpose are a natural generalization of the Wightman functions, which embody the statistical information about the field. They are also an extension of the reduced density matrices employed in many-body theory. Finally, they are closely linked with measurements.

(ii) COHERENT STATES   We have become acquainted with a new set of states, the coherent states, that have many remarkable properties. Besides being naturally adapted to the description of coherence, these states are as close as possible to classical states, and they should therefore play an important role in the analysis of the classical limit of quantum theory. Furthermore, they form a new basis, quite different from the familiar Fock state basis, that not only provides a new viewpoint, but is also much more suitable than the Fock basis for dealing with a variety of problems. Many theoretical physicists have become so accustomed to thinking in terms of Fock states that they may get a distorted picture by sticking to Fock space in such problems.

The generalized phase space representations in terms of coherent states, of which the diagonal representation is an example, have given us new insight into the relationship between classical and quantum descriptions, and they also provide a valuable new tool.

Besides the applications described in these lectures, coherent states have also been applied in the treatment of the infrared problem in quantum electrodynamics[37] (where they allow one to go outside of Fock space), in the theory of superfluids[38], and in the theory of strong interactions[39]. The latter application illustrates the point that, if at least the low-lying excitations of a system (in this case the resonances found in elementary particle physics) form a set of equally spaced levels, so that they have some features in common with a harmonic oscillator, coherent states may play a significant role in the analysis of the system.

(iii) NONLINEAR PROBLEMS   We have discussed some of the new qualitative features due to essentially nonlinear effects. In the laser, nonlinear effects are responsible for stabilizing the output, and they also govern the competition among the modes, which has many aspects in common with the competition among populations of different biological species.

The generation and propagation of intense short pulses in resonant media also gives rise to qualitatively new nonlinear phenomena. A different type of coherence is a common feature of these phenomena, corresponding to the cooperative effects that arise from the interaction of a large number of atoms with a common radiation field. A striking illustration of the surprises that may be found in this domain is provided by self-induced transparency. The 2π hyperbolic secant pulses are also interesting as exact self-consistent solutions of the coupled Maxwell and Schrödinger equations in a simplified model. They appear to be closely related with the "solitons" that have been found to be associated with other nonlinear partial differential equations describing propagation in dispersive media.

(iv) THE LASER   We have gained some insight into the behavior of the laser. This is a model of an open system, interacting with gain and loss reservoirs, that has a relatively stable state far from thermal equilibrium. The pumping reservoirs supply energy to the system, and this may be done through incoherent processes at low frequency (e.g., by electron collisions in a He–Ne gas discharge). Above a certain threshold, a kind of phase transition takes place: from a disorganized, near-thermal type of behavior, the system goes over into a highly organized form of collective behavior, producing a high-frequency, nearly monochromatic oscillation, with its amplitude stabilized by nonlinear effects. The organization is produced through the electromagnetic field, which shows long-range phase correlations (coherence).

We have here an example of a many-body problem in which, while most modes remain close to thermal equilibrium, a few modes of collective motion become very strongly excited, showing phase correlations over macroscopic distances. Other examples are provided by superfluids and superconductors, in which the coherent behavior (long-range quantum phase correlations) appears only at low temperatures, due to the nature of the stabilizing mechanism.

Another example of an open system (exchanging both energy and matter with its surroundings[40]) having a relatively stable state far from thermal equilibrium, and showing strongly correlated behavior in a few collective modes, is a biological system, e.g., a cell or a collection of cells. Considerations of this type have led Fröhlich[41] to conjecture that long-range quantum phase correlations may exist in biological systems, giving rise to strong coherent excitation and energy storage in a few modes. He proposed a specific model leading to such behavior for a system of cells, in which longitudinal electric oscillations would perform functions similar to those of the laser field. Although such considerations seem highly speculative at present, they suggest that some of the concepts we have discussed in connection with quantum optics may come to play a significant role in biophysics.

References

1. R. H. Dicke, *Phys. Rev.* **93**, 99 (1954).
2. N. A. Kurnit, I. D. Abella, and S. R. Hartmann, *Phys. Rev. Letters* **13**, 567 (1964).
3. S. L. McCall and E. L. Hahn, *Phys. Rev. Letters* **18**, 9 (1967).
4. The total transition dipole moment is $\mathbf{p}_{ab} = p(\hat{\mathbf{x}} - i\hat{\mathbf{y}})$, and its magnitude is $\sqrt{2}p$.
5. A. R. Edmonds, *Angular Momentum in Quantum Mechanics*, Princeton University Press (1957), p. 17.
6. C. Stroud, in *Electromagnetic Interactions of Two-Level Atoms*, ed. J. H. Eberly, Univ. of Rochester (1970), p. 77.
7. A more precise description of the atomic states corresponding to this model is given in §A. 2 (Appendix).
8. J. S. Ames and F. D. Murnaghan, *Theoretical Mechanics*, Ginn and Co., Boston (1929), p. 155.
9. E. U. Condon and G. H. Shortley, *The Theory of Atomic Spectra*, Cambridge Univ. Press (1935), pp. 59–61.
10. R. H. Dicke, in *Quantum Electronics*, ed. P. Grivet and N. Bloembergen, vol. 1, Dunod, Paris (1964), p. 36. See also §A. 6 (Appendix).
11. N. A. Kurnit, I. D. Abella and S. R. Hartmann, *Phys. Rev. Letters* **13**, 567 (1964); *Phys. Rev.* **141**, 391 (1966); S. R. Hartmann, QO (See Ch. 1, Ref. 3), p. 532.
12. E. L. Hahn, *Phys. Rev.* **80**, 580 (1950).
13. For a linearly polarized wave, there would also be a "counterrotating" component that would be antiresonant; neglecting it is equivalent to the rotating-wave approximation of §6.1.
14. Cf., however, §A. 1 (d) (Appendix)
15. I. D. Abella, N. A. Kurnit, and S. R. Hartmann, *Phys. Rev.* **141**, 391 (1966). Cf. also §A. 6 (Appendix)

16. S. L. McCall and E. L. Hahn, *Phys. Rev. Letters* **18**, 908 (1967). *Phys. Rev.* **183**, 457 (1969).

17. Cf. also A. Icsevgi and W. E. Lamb, *Phys. Rev.* **185**, 517 (1969).

18. F. Bloch, *Phys. Rev.* **70**, 460 (1946).

19. M. D. Crisp, *Opt. Commun.* **1**, 59 (1969).

20. J. H. Eberly, *Phys. Rev. Letters* **22**, 760 (1969).

21. By (8.6.42), we have $\dot{w} = Np \, \dot{\varphi} \sin \varphi$, and this must be positive for small φ, since the excitation energy initially increases. Thus, we must have a positive sign in (8.6.45), corresponding to the negative square root in (8.6.44).

22. The negative sign found for $\mathscr{E}$ differs from McCall and Hahn's results due to a different choice of conventions. By (8.6.41) and (8.6.26), we conclude that $\varphi(z)$ is a constant, which we set $= 0$; we could equally well choose $\varphi = \pi$ and change $\mathscr{E}$ to $-\mathscr{E}$ in (8.6.3).

23. E. Courtens, *Phys. Rev. Letters* **21**,3 (1968).

24. W. Heitler, *The Quantum Theory of Radiation*, 3rd ed., Clarendon Press, Oxford (1954), pp. 66, 69.

25. C. K. N. Patel and R. E. Slusher, *Phys. Rev. Letters* **19**, 1019 (1967); C. K. N. Patel, *Phys. Rev. A* **1**, 979 (1970).

26. H. M. Gibbs and R. E. Slusher, *Phys. Rev. Letters* **24**, 638 (1970).

27. N. S. Shiren, *Phys. Rev. B* **2**, 2471 (1970).

28. This can also be achieved by the pulse envelope changing sign, which indeed tends to happen at the trailing edge of the pulse.

29. R. H. Dicke, in *Quantum Electronics*, ed. P. Grivet and N. Bloembergen, vol. 1, Dunod, Paris (1964), p. 36.

30. J. A. Armstrong and E. Courtens, *IEEE J. Quantum Electron.* **QE-4**, 411 (1968); F. A. Hopf and M. O. Scully, *Phys. Rev.* **179**, 399 (1969); S. L. McCall and E. L. Hahn, *Phys. Rev.* **183**, 457 (1969); A. Icsevgi and W. E. Lamb, *Phys. Rev.* **185**, 517 (1969).

31. Cf., e.g., H. M. Nussenzveig, *Causality and Dispersion Relations*, Academic Press, N. Y. (1972).

32. If an observer at A at $t = 0$ sends a light signal to B, triggering the explosion of a bomb upon arrival at B, the distance $\overline{AB}$ being equal to cT, and if a bomb is exploded at A at $t = T$, one cannot say that the "speed of propagation of the explosion" is infinite; one event is not caused by the other.

33. R. C. T. da Costa, *J. Math. Phys.* **11**, 2799 (1970).

34. E. T. Jaynes and F. W. Cummings, *Proc. IEEE* **51**, 89 (1963); J. H. Eberly, *Phys. Rev. Letters* **22**, 760 (1969); M. D. Crisp, *Phys. Rev. Letters* **22**, 820 (1969).

35. G. L. Lamb, Jr., *Rev. Mod. Phys.* **43**, 99 (1971). Cf. also T. W. Barnard, *Phys. Rev. A* **7**, 373 (1973); P. J. Caudrey, J. D. Gibbon, J. C. Eilbeck and R. K. Bullough, *Phys. Rev. Letters* **30**, 237 (1973).

36. Cf. R. C. Davidson, *Methods in Nonlinear Plasma Theory*, Academic Press, New York (1972).

37. T. W. B. Kibble, *J. Math. Phys.* **9**, 315 (1968); *Phys. Rev.* **173**, 1527 (1968); **174**, 1882 (1968); **175**, 1624 (1968).

38. F. W. Cummings and J. R. Johnston, *Phys. Rev.* **151**, 105 (1966).

39. V. Alessandrini, D. Amati, M. Le Bellac and D. Olive, *Physics Reports* **1C**, 269 (1971)

40. M. O. Scully and W. E. Lamb, *Phys. Rev.* **159**, 208 (1967).

41. H. Fröhlich, in *Theoretical Physics and Biology*, ed. M. Marois, North-Holland Publishing Co., Amsterdam (1969), p. 13; *Phys. Letters* **26A**, 402 (1968); *Intern. J. Quantum Chem.* **2**, 641 (1968); *Nature* **219**, 743 (1968).

APPENDIX

Recent Developments in Quantum Optics

IN THIS APPENDIX we review some recent advances in quantum optics, mainly in the area of coherent interaction effects.

A.1   SURVEY OF SOME RECENT EXPERIMENTAL RESULTS

We mention only a few of many interesting experimental results:

(a) Nonlinear optics

A variety of nonlinear optical effects that take place when high-intensity laser beams propagate through matter have been observed[1]. They include the coupling of light waves with various types of other excitations, parametric conversion, and self-focusing and trapping effects, leading to the formation of tiny optical filaments, with diameters of just a few wavelengths.

(b) Picosecond pulses

Ultrashort pulses with durations of the order of picoseconds (10^{-12} s), have been produced; they are the object of intensive current investigation[2].

(c) Self-induced transparency

The preliminary results by Gibbs and Slusher mentioned in § 8.6(d) have been followed by a detailed experimental verification[3] of the McCall-Hahn theory. The Hg-Rb system they employed had several advantages over previous experiments, and it made possible a quantitative comparison with the theory. Several features of the theory were tested: (i) Nonlinear transmission through the sample (transparency); (ii) Large pulse delays; (iii) Pulse breakup of large pulses into several 2π pulses; (iv) Peak amplification, i.e., the fact that the peak of the output intensity exceeds the peak of the input intensity for certain values of the input area; (v) Pulse compression: peak amplification and pulse shortening lead to pulse compression effects; by focusing, a time compression by an order of magnitude can be achieved. Excellent agreement between the experimental data and the predictions of the McCall-Hahn theory was found in all cases. The effect of "chirping" (frequency modulation), corresponding to a possible dependence of the phase φ on t [neglected in (8.6.3)], was found to be small in these experiments.

More recently, the behavior of very short pulses, with pulse width smaller than the inhomogeneous broadening lifetime T_2^*, has also been investigated; the results are similar to those obtained for broader pulses. The effects of damping [corresponding to the lifetimes T_1, T_2' of § 8.6(a)] have also been discussed. The area theorem of § 8.6(c) is no longer valid when damping effects are appreciable.

The recent development of stable continuously operating tunable dye lasers should allow the extension of these techniques to a great variety of absorbers.

(d) New coherent atomic interaction effects

Several new cooperative atomic effects of the type discussed in Chapter 8 have been observed:

(i) TRANSIENT OPTICAL NUTATION   This is the optical analog of transient nutation in nuclear magnetic resonance[5]. If one excites the system coherently by a step-function pulse (i.e., a pulse with sharp rise time and long duration), with electric field $E = E_1$, the tipping motion (8.5.3) of the dipole moment, seen from the fixed (laboratory) frame, corresponds to a nutation at the Rabi flipping frequency

$$\Omega = [(\varkappa E_1)^2 + (\varDelta\omega)^2]^{1/2}, \tag{A.1.1}$$

where $\varkappa$ is given by (8.2.8). The oscillations in this optical ringing effect represent the alternate absorption and emission of radiation by the system. The effect was first observed[6] with the help of a pulsed laser. According to (8.2.8), the measurement of the nutation frequency (A.1.1) allows a direct determination of the transition dipole moment of optical transitions.

A simple new technique for the observation of coherent interaction effects, applicable to molecules with an electric dipole moment, was developed by Brewer and Shoemaker[7]. Instead of a pulsed laser source, as had been used in previous experiments, they employ a continuous wave laser. The initially degenerate molecular levels are split by sudden application of an electric field pulse, which tunes the molecular level splitting to the fixed laser frequency by the Stark effect; thus it is the two-level system, rather than the field, that is pulsed. The method was applied[7] to observe photon echoes as well as optical nutation.

(ii) OPTICAL FREE INDUCTION DECAY   This is the optical analog of free induction decay in nuclear magnetic resonance[8].

With reference to Fig. 8.5(c), the effect is represented by the decaying signal following the switching off of the pulse marked 1. The decay is due to coherent spontaneous emission (possibly including superradiance) as well as to the dephasing of individual dipoles due to inhomogeneous broadening. The effect was observed by the Brewer-Shoemaker technique.[9]

(iii) TWO-PHOTON SUPERRADIANCE   When a step-function Stark pulse is applied to a degenerate two-level system, there arises the possibility of a splitting into four levels, such that three of them can be connected by a two-photon process; e.g., by emission of a photon and absorption of a laser photon. Due to the coherent excitation, the effect exhibits the cooperative features associated with superradiance: the signal propagates in the forward direction (it is colinear with the laser beam) and it is proportional to N^2. The effect was observed by Shoemaker and Brewer[10], who have called it *two-photon superradiance*.

These new coherent transient effects and observation techniques should have important applications in the measurement of optical transition parameters and in the study of relaxation processes.

Significant theoretical advances have recently been made in the treatment of coherent interaction effects, particularly superradiance. The remainder of this appendix is devoted to a survey of these theoretical developments[11].

A.2   COHERENT ATOMIC STATES

It has been shown by Radcliffe[12] and by Arecchi *et al*[13], that a free system of N two-level atoms can be described in terms of a new set of basis states, the *coherent atomic states*, having properties that show a far-reaching analogy with those of the coherent states for the electromagnetic field described in Chapter 3.

This analogy is not accidental. Physically, it is related with the observation, already made in § 8.7(ii), that the energy levels of a system of two-level atoms are equally spaced up to the highest (fully excited) level, so that the spectrum is analogous to that of a harmonic oscillator, the crucial difference being, of course, that it terminates at a maximum excitation. For low-lying excitations, the two systems should show similar properties, and the harmonic oscillator should correspond to a limiting case as the number of levels goes to infinity. In mathematical terms this is related with the concept of "group contraction", leading from the angular momentum operator algebra to the harmonic oscillator operator algebra (see below).

Let us go back to the angular-momentum-like operators $\mathscr{R}_l$ ($l = 1, 2, 3$), $\mathscr{R}_\pm$, $\mathscr{R}^2$ of § 8.3, and to the corresponding Dicke states $|r, m\rangle$ defined by (8.3.16) and (8.3.17). As we have seen in (8.3.19), the Dicke states are still degenerate, and one would need additional quantum numbers α to get a complete characterization of a state $|r, m; \alpha\rangle$. These additional quantum numbers required to get a complete set can be obtained[13] by adding to $\mathscr{R}^2$ and $\mathscr{R}_3$ some operators of the permutation group of N objects. For the remainder of this section, we stay within the subspace of Hilbert space corresponding to a given cooperation number r and invariant under the above-mentioned permutation operators. For simplicity, we omit the corresponding quantum numbers r and α from the labeling of the states, employing the notation $|m\rangle$ instead of $|r, m; \alpha\rangle$ for a Dicke state.

The Dicke states $|m\rangle$ $(-r \leq m \leq r)$ can be generated from the ground state $|-r\rangle$ by repeated application of the raising operator $\mathscr{R}_+$ (cf. § 8.3) in the following manner:

$$|m\rangle = \left[\frac{(r-m)!}{(2r)!\,(r+m)!}\right]^{\frac{1}{2}} \mathscr{R}_+^{r+m}\,|-r\rangle, \quad \langle m\,|\,m\rangle = 1,$$

$$m = -r, \ldots, r \quad [(3.1.12)], \tag{A.2.1}$$

where

$$\mathscr{R}_-\,|-r\rangle = 0 \quad [(3.1.11)]. \tag{A.2.2}$$

From now on, to stress the analogy with the coherent states for the field, we place in square brackets the number of the corresponding equation for the field; thus, (A.2.1) and (A.2.2) are analogous to (3.1.12) and (3.1.11), respectively: *The Dicke states $|m\rangle$ are the analogues of the Fock states $|n\rangle$.*

The *coherent atomic state* $|\tau\rangle$ is defined by

$$|\tau\rangle = (1 + |\tau|^2)^{-r} \exp(\tau\mathscr{R}_+)\,|-r\rangle$$

$$= (1 + |\tau|^2)^{-r} \sum_{m=-r}^{r} \binom{2r}{r+m}^{\frac{1}{2}} \tau^{r+m}\,|m\rangle,$$

$$\langle \tau\,|\,\tau\rangle = 1, \quad [(3.2.1), (3.1.24)], \tag{A.2.3}$$

where τ is an arbitrary complex number, and $\binom{n}{j}$ is a binomial coefficient. In particular, for $\tau = 0$, we recover the ground state $|-r\rangle$. We have, by (A.2.3),

$$\langle \tau'\,|\,\tau\rangle = \frac{(1 + \tau'^*\tau)^{2r}}{(1 + |\tau'|^2)^r\,(1 + |\tau|^2)^r} \quad [(3.2.27)], \tag{A.2.4}$$

corresponding to the *nonorthogonality* of different coherent states.

Let us evaluate the expectation value of the operators $\mathscr{R}_l$ in the coherent state $|\tau\rangle$. We have, by (A.2.3),

$$\langle\tau|\,\mathscr{R}_3\,|\tau\rangle = (1 + |\tau|^2)^{-2r} \sum_{j=0}^{2r} (j - r)\binom{2r}{j}|\tau|^{2j}$$

$$= -r + (1 + \varrho)^{-2r}\,\varrho\,\frac{d}{d\varrho}\,(1 + \varrho)^{2r}, \quad \varrho \equiv |\tau|^2,$$

so that

$$\langle\tau|\,\mathscr{R}_3\,|\tau\rangle = -r + \frac{2r\,|\tau|^2}{1 + |\tau|^2}. \tag{A.2.5}$$

Similarly, since (up to a phase factor)

$$\mathscr{R}_+\,|r, m\rangle = \sqrt{(r - m)\,(r + m + 1)}\,|r, m + 1\rangle, \tag{A.2.6}$$

we find

$$\langle\tau|\,\mathscr{R}_+\,|\tau\rangle = (1 + |\tau|^2)^{-2r} \sum_{j=0}^{2r} (j/\tau)\binom{2r}{j}|\tau|^{2j}$$

$$= (1 + |\tau|^2)^{-2r}\,\frac{d}{d\tau}\,(1 + |\tau|^2)^{2r},$$

so that

$$\langle\tau|\,\mathscr{R}_+\,|\tau\rangle = \frac{2r\tau^*}{1 + |\tau|^2}, \quad \langle\tau|\,\mathscr{R}_-\,|\tau\rangle = \frac{2r\tau}{1 + |\tau|^2}. \tag{A.2.7}$$

These results acquire a transparent physical significance when we parametrize the coherent states by

$$\tau = \tan\left(\frac{\theta}{2}\right)e^{-i\varphi} \quad (0 \leqq \theta < \pi, \quad 0 \leqq \varphi < 2\pi). \tag{A.2.8}$$

We may then rewrite (A.2.3) as

$$|\theta, \varphi\rangle = |\tau\rangle = \left(\cos\frac{\theta}{2}\right)^{2r} \exp\left[\tan\left(\frac{\theta}{2}\right)e^{-i\varphi}\mathscr{R}_+\right]|-r\rangle, \tag{A.2.9}$$

and (A.2.5), (A.2.7) become

$$\langle\theta, \varphi|\,\mathscr{R}_3\,|\theta, \varphi\rangle = -r\cos\theta, \tag{A.2.10}$$

$$\langle\theta, \varphi|\,\mathscr{R}_\pm\,|\theta, \varphi\rangle = r\sin\theta\exp(\pm i\varphi). \tag{A.2.11}$$

In terms of the vector operator $\mathscr{R} = (\mathscr{R}_1, \mathscr{R}_2, \mathscr{R}_3)$, we get

$$\langle\theta, \varphi|\,\mathscr{R}\,|\theta, \varphi\rangle = \mathbf{r} = (r, \theta, \varphi), \tag{A.2.12}$$

where $\mathbf{r}$ is the position vector on the surface of a sphere of radius r in the direction (θ, φ), and the angle θ is measured *from the south pole* instead of the north pole (as is usual for the colatitude). We recognize this (apart from

14*

a normalization factor for the radius) as the *Bloch sphere* representation employed throughout Chapter 8.

Actually, the coherent atomic state $|\theta, \varphi\rangle$ can be obtained[13] by the rotation of the ground state $|-r\rangle \equiv |\theta = 0\rangle$ (which corresponds to the south pole),

$$|\theta, \varphi\rangle = R_{\theta,\varphi}|-r\rangle, \quad [(3.2.7)] \tag{A.2.13}$$

where $R_{\theta,\varphi}$ is the rotation operator which takes the direction of the south pole into the direction (θ, φ). In fact, by "disentangling" techniques, one can show[13] that

$$R_{\theta,\varphi} = \exp\left[\frac{\theta}{2}(e^{-i\varphi}\mathscr{R}_+ - e^{i\varphi}\mathscr{R}_-)\right]$$

$$= \exp\left(\tau\mathscr{R}_+\right)\exp\left[\ln(1 + |\tau|^2)\,\mathscr{R}_3\right]\exp\left(-\tau^*\mathscr{R}_-\right) \quad [(3.2.6)], \tag{A.2.14}$$

and (A.2.13) is then readily identified with (A.2.3). Thus, just as the coherent field states correspond to displaced harmonic oscillator ground states [translated in the oscillator phase plane; cf. § 3.2(a)], *the coherent atomic states correspond to angular momentum ground states rotated on the Bloch sphere*; Arecchi *et al.* propose to call them "Bloch states".

Let $\hat{\mathbf{r}}$ be a unit vector in the direction (θ, φ) [cf. (A.2.12)]. Then, (A.2.13) implies

$$(\mathscr{R} \cdot \hat{\mathbf{r}})|\hat{\mathbf{r}}\rangle = r|\hat{\mathbf{r}}\rangle, \quad |\hat{\mathbf{r}}\rangle = |\theta, \varphi\rangle, \tag{A.2.15}$$

i.e., *the coherent atomic state* $|\hat{\mathbf{r}}\rangle = |\theta, \varphi\rangle$ *is an eigenstate of the "pseudo-angular momentum in the direction* $\hat{\mathbf{r}}$*"* (with pseudo-angular momentum $\mathscr{R}$ quantized "along the direction $\hat{\mathbf{r}}$").

The rotated $\mathscr{R}_l$ operators,

$$\mathscr{R}'_l = R_{\theta,\varphi}\mathscr{R}_l R_{\theta,\varphi}^{-1} \quad (l = 1, 2, 3) \tag{A.2.16}$$

still satisfy the commutation relation $[\mathscr{R}'_1, \mathscr{R}'_2] = i\mathscr{R}'_3$, which leads[14] to the Heisenberg uncertainty relation

$$\langle(\Delta\mathscr{R}'_1)^2\rangle \langle(\Delta\mathscr{R}'_2)^2\rangle \geq \tfrac{1}{4}\langle(\mathscr{R}'_3)^2\rangle \quad [(3.2.23)]. \tag{A.2.17}$$

The equality sign holds if the expectation value is taken in the state $|\theta, \varphi\rangle$: *the coherent atomic states are minimum-uncertainty states* [cf. § 3.2(b)].

The parameter τ given by (A.2.8) has a simple geometric interpretation. Consider a Bloch sphere of unit diameter (radius $= \tfrac{1}{2}$) and identify its tangent plane at the south pole with the complex plane. Each point P on the surface of this sphere may be associated with the coherent atomic state defined by the corresponding direction (θ, φ). If we now project P on the tangent plane, using the north pole as the center of projection, it is readily verified from (A.2.8) that the image on the tangent plane is the complex number τ^*. This transformation corresponds to the well known *stereographic projection*[15], which establishes a one-to-one conformal correspondence between the (extended) complex plane and the Riemann sphere. The factor $(1 + |\tau|^2)^{-1}$ which appears in the above formulae is the *magnification*, i.e., the ratio between corresponding line elements on the sphere and in the plane.

It follows from (A.2.4) that

$$|\langle \tau' \mid \tau \rangle| = [1 - \chi^2(\tau'^*, \tau^*)]^r, \qquad (A.2.18)$$

where

$$\chi(z, z') = \frac{|z - z'|}{(1 + |z|^2)^{1/2}(1 + |z'|^2)^{1/2}} \qquad (A.2.19)$$

is called the *chordal distance*[15] between the points r and u'; it is the length of the chord connecting the corresponding spherical images. In terms of $\hat{\mathbf{r}} = (\theta, \varphi)$ and $\hat{\mathbf{r}}' = (\theta', \varphi')$, we have

$$\langle \theta', \varphi' \mid \theta, \varphi \rangle \equiv \langle \hat{\mathbf{r}}' \mid \hat{\mathbf{r}} \rangle$$

$$= \left\{ \cos\frac{\theta'}{2} \cos\frac{\theta}{2} + \sin\frac{\theta'}{2} \sin\frac{\theta}{2} \exp\left[i(\varphi' - \varphi)\right] \right\}^{2r}, \qquad (A.2.20)$$

so that

$$|\langle \hat{\mathbf{r}}' \mid \hat{\mathbf{r}} \rangle| = \left(\frac{1 + \hat{\mathbf{r}} \cdot \hat{\mathbf{r}}'}{2}\right)^r = (1 - \tfrac{1}{4} | \hat{\mathbf{r}} - \hat{\mathbf{r}}'|^2)^r = \left(\cos\frac{\Theta}{2}\right)^{2r} \ [(3.2.28)],$$

$$(A.2.21)$$

where Θ is the angle between the directions (θ, φ) and (θ', φ'). Thus the overlap between two coherent atomic states decreases as their images move further apart on the Bloch sphere; two states with diametrically opposite spherical images are orthogonal.

The coherent atomic states also form an overcomplete set [cf. § 3.2(d)] in the finite-dimensional Hilbert subspace with fixed cooperation number r that we are considering. To see this, we evaluate the "completeness sum"

$$\int |\hat{\mathbf{r}}\rangle \langle \hat{\mathbf{r}}| \, d\Omega = \int |\theta, \varphi\rangle \langle \theta, \varphi| \sin\theta \, d\theta \, d\varphi$$

$$= 4\pi \sum_{m=-r}^{r} \binom{2r}{r+m} \int_0^\pi \left(\sin\frac{\theta}{2}\right)^{2r+2m+1} \left(\cos\frac{\theta}{2}\right)^{2r-2m+1} d\theta \, |m\rangle \langle m|$$

$$= \frac{4\pi}{2r+1} \sum_{m=-r}^{r} |m\rangle \langle m| = \frac{4\pi}{2r+1}, \qquad (A.2.22)$$

where we have employed (A.2.9) and the completeness of the Dicke states in the above-mentioned Hilbert subspace. Thus, we finally get the *resolution of unity*

$$\frac{(2r+1)}{4\pi} \int |\theta, \varphi\rangle \langle\theta, \varphi| \sin\theta \, d\theta \, d\varphi = (2r+1) \int |\hat{\mathbf{r}}\rangle \langle\hat{\mathbf{r}}| \frac{d\Omega}{4\pi} = 1$$

$$[(3.2.31)]. \quad \text{(A.2.23)}$$

It follows that, for an arbitrary state vector $|f\rangle$ in this subspace,

$$|f\rangle = \sum_{m=-r}^{r} f_m |m\rangle \quad [(3.3.2)], \qquad \text{(A.2.24)}$$

we have the coherent state expansion

$$|f\rangle = (2r+1) \int \langle \hat{\mathbf{r}} | f\rangle | \hat{\mathbf{r}}\rangle \frac{d\Omega}{4\pi}$$

$$= (2r+1) \int (1 + |\tau|^2)^{-r} f(\tau^*) | \hat{\mathbf{r}}\rangle \frac{d\Omega}{4\pi} \quad [(3.3.7)], \quad \text{(A.2.25)}$$

where, according to (A.2.3) and (A.2.24),

$$f(\tau^*) = (1 + |\tau|^2)^r \langle \hat{\mathbf{r}} | f\rangle = \sum_{m=-r}^{r} \binom{2r}{r+m}^{\frac{1}{2}} f_m(\tau^*)^{r+m}$$

$$[(3.3.4)] \quad \text{(A.2.26)}$$

is a *polynomial* in τ^* of degree $2r$.

Similarly, for an arbitrary operator F acting on our subspace, we can immediately write down a nondiagonal representation analogous to (3.3.18). However, in view of the fact that F is completely defined by its $(2r+1)^2$ matrix elements $\langle m|F|m'\rangle$ in the Dicke basis, one can show[13] that there always exists a *diagonal coherent-state representation* for F,

$$F = \int F(\theta, \varphi) |\theta, \varphi\rangle \langle\theta, \varphi| \, d\Omega \quad [(3.4.2)], \qquad \text{(A.2.27)}$$

where the expansion coefficients may be expressed in terms of the matrix elements $\langle m|F|m'\rangle$. In particular, the expansion coefficients associated with a density operator define a *quasiprobability distribution* on the Bloch sphere [cf. § 3.4(e)].

The mathematical basis underlying the parallel between the above properties of coherent atomic states and the corresponding properties of coherent field states discussed in Chapter 3 is the process of *group contraction*[16]. This is a method whereby, starting from a given Lie group, one obtains another, non-isomorphic one by a limiting procedure; e.g., the Galilei group may be obtained from the Lorentz group by group contraction, letting the velocity of light approach infinity. In the present case, the limiting process consists in letting the radius r of the Bloch sphere go to infinity, in such a way that small rotations on the sphere go over into translations in the tangent plane at the south pole, which corresponds to the phase plane of the harmonic oscillator. This leads from the angular momentum algebra to the harmonic oscillator algebra.

Setting

$$r = \tfrac{1}{2}\varepsilon^{-2}, \qquad \varepsilon \to 0, \tag{A.2.28}$$

one finds[13] that the operators $\varepsilon\mathcal{R}_{+}$, $\varepsilon\mathcal{R}_{-}$ and $\mathcal{R}_{3} + \tfrac{1}{2}\varepsilon^{-2}$ go over into the harmonic oscillator operators a^{+}, a and $a^{+}a$, respectively. The eigenvalue v of (3.1.10) is the limit of $\tfrac{1}{2}(\theta/\varepsilon)\,e^{-i\varphi}$, and $r + m$ corresponds to n; the Dicke state $|m\rangle$ approaches the Fock state $|n\rangle$, and the coherent atomic state $|\theta, \varphi\rangle$ approaches the coherent field state $|v\rangle$. As an illustration of these results, we note that (A.2.3) goes over into

$$\lim_{\varepsilon \to 0} (1 + \varepsilon^{2}\,|v|^{2})^{-\varepsilon^{-2}/2}\,e^{va^{+}}\,|0\rangle$$

$$= \exp\left(-\tfrac{1}{2}\,|v|^{2} + va^{+}\right)|0\rangle,$$

which, by (3.2.1), indeed corresponds to $|v\rangle$. Similarly, one can verify that the other results given above go over into the corresponding equations of Chapter 3 given in square brackets.

Let us now consider the interaction of the coherent atomic states with the electromagnetic field. The approximations are the same as in Chapter 8, i.e., we make the dipole and rotating wave approximations; furthermore, we assume either that the dimensions of the container are much smaller than the wavelength or else that the atoms interact only with a single mode of the electromagnetic field, corresponding to a given direction of propagation $\mathbf{k}$. As shown in § 8.4(c), the latter case can be reduced to the former by introducing the Dicke operators $\mathcal{R}_{\mathbf{k}}^{\pm}$ of (8.4.25). To what extent this is a realistic approximation will be discussed below in § A.6.

We have seen in § 4.1(b) that the field produced by a classical (c-number) current distribution is in a coherent state. The counterpart of this property is the following result[13]: *An atomic system of small dimension which is initially in a coherent atomic state (in particular in the ground state), and which is acted upon by a classical (c-number) field, remains in a coherent atomic state.* To see this we note that the interaction hamiltonian in the interaction picture is of the form [cf. (8.3.3)]

$$H_{I} = 2pE(\mathcal{R}_{1}\sin\varphi' - \mathcal{R}_{2}\cos\varphi'), \tag{A.2.29}$$

where E is a c-number. The Schrödinger equation (2.2.11) then yields, for an initial state $|\theta, \varphi\rangle$ at time t,

$$|t + \Delta t\rangle = [1 - i2pE\,\Delta t(\mathcal{R}_1 \sin \varphi' - \mathcal{R}_2 \cos \varphi')]\,|\theta, \varphi\rangle. \qquad \text{(A.2.30)}$$

The operator within square brackets represents an infinitesimal rotation by an angle $\Delta\theta = 2pE\,\Delta t$ (tipping angle) around an axis $(\sin \varphi', -\cos \varphi', 0)$. According to (A.2.13), this preserves the coherent character of the atomic state.

Finally, let us compute the radiation rate from a coherent atomic state in the approximation of §.8.4, i.e., in first order perturbation theory. From (8.4.2) and (8.4.5), we find, for a state $|\tau\rangle$,

$$I/I_0 = \sum_{m=-r}^{r} (r + m)(r - m + 1)\,|\langle m\,|\,\tau\rangle|^2. \qquad \text{(A.2.31)}$$

By the same method employed in the evaluation of (A.2.5) and with the same notation $\varrho \equiv |\tau|^2$, we get, with the help of (A.2.3),

$$I/I_0 = (1 + \varrho)^{-2r} \sum_{j=0}^{2r} j[2r - (j - 1)]\binom{2r}{j}\varrho^j$$

$$= (1 + \varrho)^{-2r}\left[2r\varrho\,\frac{d}{d\varrho}(1 + \varrho)^{2r} - \varrho^2\,\frac{d^2}{d\varrho^2}(1 + \varrho)^{2r}\right]$$

$$= 4r\,\frac{\varrho}{1 + \varrho} - 2r(2r - 1)\,\frac{\varrho^2}{(1 + \varrho)^2}.$$

Substituting $\varrho = |\tau|^2 = \tan^2 \dfrac{\theta}{2}$ [cf. (A.2.8)], this becomes

$$I(\theta, \varphi)/I_0 = r^2 \sin^2 \theta + 2r \sin^4 \frac{\theta}{2} \qquad \text{(A.2.32)}$$

for the radiation from the coherent atomic state $|\theta, \varphi\rangle$.

On the other hand, by (8.4.2), the radiation rate from the Dicke state having the same energy expectation value (A.2.10) is

$$I(r, m = -r \cos \theta)/I_0 = r^2 \sin^2 \theta + 2r \sin^2 \frac{\theta}{2}. \qquad \text{(A.2.33)}$$

Comparing this with (A.2.32), we see that *the radiation rates from a Dicke state and from a coherent atomic state with the same energy expectation value are practically identical for $r \gg 1$.* In particular, both are superradiant for $\theta \approx \pi/2$.

In spite of the similar radiation rate, however, there is a considerable physical difference between the two kinds of states. To see this, let us compute the expectation value of the total dipole moment operator, which, in the Heisenberg picture, is given by [cf. (8.4.18), (8.4.24)]

$$\mathscr{P}_\perp = \mathbf{p}e^{i\omega t}\mathscr{R}_+ + \mathbf{p}^*e^{-i\omega t}\mathscr{R}_-. \tag{A.2.34}$$

For a Dicke state, we obviously have

$$\langle r, m | \mathscr{P}_\perp | r, m \rangle = 0, \tag{A.2.35}$$

whereas, for the coherent atomic state $|\theta, \varphi\rangle$, (A.2.34) and (A.2.11) yield

$$\langle \theta, \varphi | \mathscr{P}_\perp | \theta, \varphi \rangle = r \sin \theta (\mathbf{p}e^{i\omega t + i\varphi} + \mathbf{p}^*e^{-i\omega t - i\varphi}). \tag{A.2.36}$$

Thus, the expectation value of the dipole moment vanishes in a Dicke state, whereas, *in a coherent atomic state, there is a nonvanishing oscillating average dipole moment,* for which the classical radiation rate would be

$$I_{\text{class}}/I_0 = r^2 \sin^2 \theta. \tag{A.2.37}$$

This differs from (A.2.32) only for θ close to π. For a fully inverted state there would be no classical emission, whereas first-order perturbation theory gives a quantum radiation rate $2rI_0$ (cf. also § A.6).

We see, therefore, that Dicke states are "nonclassical", in the same sense as Fock states, which also have zero expectation value for the electric fields [§ 7.3(b)], whereas coherent atomic states, just like coherent field states, are the closest quantum analogues to classical states (in the atomic case, the analogy is with macroscopic dipoles).

The difference between superradiant Dicke states and superradiant coherent atomic states can also be expressed[17] in terms of the presence or absence of quantum correlations among the atoms. A superradiant coherent atomic state corresponds to a direct product of single-atom superposition states, of the type [cf. (8.3.6)]

$$\prod_{j=1}^{N} (a_j |+\rangle_j + b_j |-\rangle_j), \quad |a_j|^2 + |b_j|^2 = 1, \tag{A.2.38}$$

with $|a_j| = |b_j| = 2^{-\frac{1}{2}}$, and with the same phase difference between a_j and b_j for all j. This is an *uncorrelated state*. On the other hand, a superradiant Dicke state corresponds to a symmetric superposition of states of the type (8.3.6) (an example is the state $|1, 0\rangle$ in Fig. 8.1), which shows strong quantum mechanical atom-atom correlations.

As shown by Senitzky[17] through a first-order perturbative calculation (cf. also § A.6), the radiation emitted by the superradiant coherent atomic state is coherent (having a well-defined phase), whereas radiation from a superradiant Dicke state is incoherent. Thus, the origin of the enhanced radiation rate proportional to N^2 is quite different in the two cases: in the coherent atomic case, the enhancement factor corresponds to that for a set of classical dipole oscillators that are all oscillating in phase, whereas in the Dicke case it arises from the fact that the set of all atoms behaves like a single quantum system.

This complementary relation between quantum atomic correlations and coherence, with lack of correlation implying maximum coherence, and vice-versa, is similar to that found in connection with the factorization property for the degree of coherence of a coherent field in § 2.4(b); it is also related with the uncertainty principle[18].

The above discussion renders more precise the qualitative considerations of § 8.4(b) about the classical vector model, which are now seen to refer to coherent atomic states.

One may well ask which of the two kinds of superradiant states is actually observed experimentally. If the atoms are coherently pumped from the ground state by a strong short pulse of coherent light, that may be treated like a classical source acting for a short time, it is clear from (A.2.30) that the result is a coherent atomic state (the radiation from which has a well-defined phase relation with the incident pulse); it seems much harder to find an excitation method that would enable one to prepare a superradiant Dicke state[19]. The character of the initial state may change as a result of its dynamical evolution and interaction with the radiation field; for further discussion of this point, see § A.6.

A.3   FIELD QUANTIZATION IN DICKE'S MODEL

The treatment of superradiance in § 8.4 was semiclassical and it was based on first-order perturbation theory. As a first step toward the removal of these restrictions, let us quantize the electromagnetic field in Dicke's model. Spontaneous emission is thereby automatically included.

We assume, for simplicity, that the atoms interact with a single cavity mode of frequency ω. In the rotating wave approximation, the hamiltonian of the system is

$$H = H_0 + H_I, \tag{A.3.1}$$

where [cf. (8.3.3) and (8.4.24)]

$$H_0 = H_F + H_A = \omega a^+ a + \omega_0 \mathcal{R}_3, \tag{A.3.2}$$

$$H_I = H_{AF} = K(a\mathcal{R}_+ + a^+ \mathcal{R}_-). \tag{A.3.3}$$

In (A.3.2) and (A.3.3), a^+ and a are creation and annihilation operators associated with the mode under consideration [cf. (2.1.7), (2.1.8)], and the coupling constant K is proportional to p as well as to $V^{-\frac{1}{2}}$, where V is the cavity volume [cf. (2.1.2), (2.1.9)]; we have also taken $\hbar = 1$.

Measuring the energy in units of the frequency ω, and taking the resonance case $\omega = \omega_0$, for simplicity, we get

$$H = \underbrace{a^+a + \mathcal{R}_3}_{H_0} + \underbrace{\varkappa(a\mathcal{R}_+ + a^+\mathcal{R}_-)}_{H_I}. \tag{A.3.4}$$

It is interesting to note[20] that the same hamiltonian, with a different interpretation, also describes an apparently very different system, related with the processes of parametric amplification and frequency conversion in nonlinear optics. To see this, we employ the Schwinger theory[21] of angular momentum, which may be thought of as building up an arbitrary angular momentum out of spin $\frac{1}{2}$ units.

It was shown by Schwinger that any angular momentum can be represented in terms of a pair of boson operators a_1 and a_2, with the usual commutation rules

$$[a_i, a_k^+] = \delta_{i,k} \quad (i, k = 1, 2), \tag{A.3.5}$$

in the following way

$$\mathcal{R}_+ = a_2^+ a_1, \quad \mathcal{R}_- = a_1^+ a_2, \quad \mathcal{R}_3 = \tfrac{1}{2}(a_2^+ a_2 - a_1^+ a_1). \tag{A.3.6}$$

It is readily seen that these operators obey the correct commutation rules,

$$[\mathcal{R}_+, \mathcal{R}_-] = 2\mathcal{R}_3, \quad [\mathcal{R}_3, \mathcal{R}_\pm] = \pm\mathcal{R}_\pm. \tag{A.3.7}$$

The operator (8.3.15) becomes

$$\mathcal{R}^2 = \hat{\mathcal{R}}(\hat{\mathcal{R}} + 1), \quad \hat{\mathcal{R}} = \tfrac{1}{2}(a_2^+ a_2 + a_1^+ a_1). \tag{A.3.8}$$

Substituting (A.3.6) in (A.3.4), we find

$$H = \omega_2 a_2^+ a_2 + \omega_1 a_1^+ a_1 + \omega a^+ a$$
$$+ \varkappa(a_2^+ a_1 a + a^+ a_1^+ a_2), \tag{A.3.9}$$

where, in our units and with our choice for the zero level of energy, $\omega_2 = \frac{1}{2}$, $\omega_1 = -\frac{1}{2}$ and $\omega = 1$.

The hamiltonian (A.3.9) describes a trilinear interaction of three boson fields, that may represent several processes of interest in nonlinear optics.[22] The interaction hamiltonian contains the processes $\omega_2 \rightleftarrows \omega_1 + \omega$, which are usually interpreted in terms of photons of frequency ω_2 that are coupled, in a nonlinear medium, to photons of frequency ω_1 and photons or phonons (or other boson excitations) of frequency ω. Depending on the situation, we may then have a parametric amplifier or oscillator, with up or down conversion of the frequency; if ω is associated with phonons, we have stimulated Raman (optical phonon) or Brillouin (acoustical phonon) scattering. These processes have been discussed[22] in the semiclassical approximation, in which the field operators are replaced by their expectation values.

With the above reinterpretation, and denoting by $|n_2, n_1\rangle$ the eigenstates

$$a_i^+ a_i |n_2, n_1\rangle = n_i |n_2, n_1\rangle \quad (i = 1, 2), \tag{A.3.10}$$

we have the correspondence [cf. (A.3.6), (A.3.8)]

$$|n_2, n_1\rangle \rightleftarrows |r, m\rangle, \tag{A.3.11}$$

where

$$m = \tfrac{1}{2}(n_2 - n_1), \quad r = \tfrac{1}{2}(n_2 + n_1), \tag{A.3.12}$$

associated with a process in which the effective populations of ω_2 and ω_1 are n_2 and n_1, respectively.

A Dicke superradiant state would correspond to $n_2 \approx n_1 \sim N/2$. In the discussion of spontaneous emission for the hamiltonian (A.3.4), the initial state of the field is the photon vacuum, i.e., the eigenstate with $n = 0$ of $a^+ a$. Due to the symmetry between the bosons ω_1 and ω in (A.3.9), the set $\{n_1 = r, n_2 = r, n = 0\}$ is essentially equivalent to $\{n_1 = 0, n_2 = r, n = r\}$. Thus, the process of spontaneous emission from a superradiant state is analogous (in the above sense) to the stimulated emission from a fully excited initial state, in which there is already present initially a number of photons equal to the effective population of the excited level. This gives some additional insight into the nature of superradiant emission. Although it would not be appropriate to call it "stimulated emission", because the atoms radiate into the vacuum (there is no external field present), the common interaction of all the atoms with the radiation field they produce gives rise to a kind of "self-stimulation"[23].

A.4   STATIONARY STATES

We now discuss the stationary states of the total system (field + atoms), within the interpretation (A.3.4) [the results can be immediately reinterpreted in the sense of (A.3.9)]. The state space for the system is

$$\mathscr{H}_F \otimes \mathscr{H}_N, \tag{A.4.1}$$

where $\mathscr{H}_F$ is the Hilbert space associated with the states of the field and

$$\mathscr{H}_N = \mathbb{C}^2(1) \otimes \mathbb{C}^2(2) \otimes \cdots \otimes \mathbb{C}^2(N) \tag{A.4.2}$$

is the Hilbert space associated with the N atoms: each term in the tensor product (A.4.2) can be represented by a complex 2×2 matrix, corresponding to a spin $\tfrac{1}{2}$ system.

The stationary states have been discussed both in terms of the Fock representation[24] for $\mathscr{H}_F$ and in terms of the coherent-state representation[25,26].

It follows from (A.3.2) and (A.3.4) that

$$[\mathscr{R}^2, H_I] = [\mathscr{R}^2, H_A] = [\mathscr{R}^2, H] = 0, \qquad (A.4.3)$$

so that the stationary states can be taken as eigenstates of $\mathscr{R}^2$ with cooperation number r.

An important observation is that the unperturbed hamiltonian in (A.3.4) commutes with the full hamiltonian,

$$[H_0, H_I] = [H_0, H] = 0, \qquad (A.4.4)$$

as may readily be verified. The stationary eigenstates of H may therefore be taken as simultaneous eigenstates of $\mathscr{R}^2$, H_0 and H. We denote them by $|r, c, k\rangle$, with

$$\mathscr{R}^2 |r, c, k\rangle = r(r + 1) |r, c, k\rangle, \qquad (A.4.5)$$

$$H_0 |r, c, k\rangle = c |r, c, k\rangle, \qquad (A.4.6)$$

$$H |r, c, k\rangle = E_{r,c,k} |r, c, k\rangle, \qquad (A.4.7)$$

where $E_{r,c,k}$ are the eigenvalues of the energy of the complete system.

The unperturbed (noninteracting) system has eigenstates of the form $||n \rangle\rangle |r, m\rangle$, where $||n\rangle\rangle$ denotes an n-photon state of the field, with

$$H_0 ||n\rangle\rangle |r, m\rangle = (n + m) ||n\rangle\rangle |r, m\rangle, \qquad (A.4.8)$$

i.e., $c = n + m$. According to (8.3.18), m ranges from $-r$ to r in the state (A.4.5), so that, if we expand $|r, c, k\rangle$ in terms of the unperturbed eigenstates (A.4.8), the number of photons n can only range from $c - r$ to $c + r$,

$$|r, c, k\rangle = \sum_{n=c-r}^{c+r} A_n^{(r,c,k)} ||n\rangle\rangle |r, c - n\rangle. \qquad (A.4.9)$$

The exact eigenstates are therefore *finite* superpositions of the unperturbed ones, and the problem is reduced to evaluating the expansion coefficients $A_n^{(r,c,k)}$.

For example, if we consider a single atom interacting with the field, i.e., $N = 1$, we must have $r = \frac{1}{2}$ and (A.4.9) becomes (c is half-integer in this case)

$$|\tfrac{1}{2}, c, k\rangle = A_{c-\frac{1}{2}} ||c - \tfrac{1}{2}\rangle\rangle |\tfrac{1}{2}, \tfrac{1}{2}\rangle + A_{c+\frac{1}{2}} ||c + \tfrac{1}{2}\rangle\rangle |\tfrac{1}{2}, -\tfrac{1}{2}\rangle, \qquad (A.4.10)$$

which is a superposition of the state with $c - \frac{1}{2}$ photons and the atom in the upper level with the state with $c + \frac{1}{2}$ photons and the atom in the lower level.

According to (A.2.6), the interaction hamiltonian H_I in (A.3.4) gives rise to transitions from $||n\rangle\rangle |r, m\rangle$ only to the adjacent states $||n \pm 1\rangle\rangle |r, m \mp 1\rangle$. Thus, the matrix elements

$$\langle\langle n|| \langle r, m| H |r', m'\rangle ||n'\rangle\rangle \qquad (A.4.11)$$

define a matrix with the following properties:

For each given value of c in (A.4.6), the matrix (A.4.11) has dimension 2^N, corresponding to the different possible values of (r, m). Each value of r (r varies from $N/2$ to 0 or to $\frac{1}{2}$, depending on whether N is even or odd) gives rise to $D_{(N/2)-r}$ identical blocks aligned along the main diagonal, where $D_{(N/2)-r}$ is given by (8.3.19), and each block has dimension $2r + 1$, corresponding to the possible values of m. The number and dimensions of the blocks correspond to the irreducible representations of the group $SU(2)$. If we now let c range over its possible values (for given r, c ranges from $-r$ to ∞), we get infinitely many blocks, each of dimension 2^N.

For a given r and c, in view of the above properties of the interaction matrix elements, (A.4.11) is a *tridiagonal matrix*, i.e., the only nonvanishing matrix elements are along the main diagonal (all diagonal elements are equal to c) and those immediately adjacent to it on either side.

Correspondingly, substituting (A.4.9) in the eigenvalue equation (A.4.7), we find the finite difference equation

$$-\varkappa \alpha_{r,c,n} A_{n-1}^{(r,c,k)} + (c - E_{r,c,k}) A_n^{(r,c,k)}$$

where
$$-\varkappa \alpha_{r,c,n+1} A_{n+1}^{(r,c,k)} = 0, \tag{A.4.12}$$

$$\alpha_{r,c,n} = \sqrt{n[r(r + 1) - (c - n)(c - n + 1)]}. \tag{A.4.13}$$

The exact solution of this equation can be explicitly written down[24]. However, it takes a quite complicated form for large values of r and/or c. Let us briefly describe some of the main properties of the solution.

For given r and c, the interaction splits the degenerate eigenvalue c into (in general) $2r + 1$ eigenvalues $E_{r,c,k}$, symmetrically displaced about the unperturbed value c,

$$E_{r,c,k} - c = \varkappa \, \varepsilon_{r,c,k}, \qquad \varepsilon_{r,c,-k} = -\varepsilon_{r,c,k},$$
$$k = -r, ..., r, \tag{A.4.14}$$

where $k = -r$ denotes the ground state. The corresponding eigenstates may be represented graphically by plotting the coefficients $A_n^{(r,c,k)}$ of the associated expansion (A.4.9) as a function of n. Physically, $A_n^{(r,c,k)}$ represents the amplitude to find n photons in the field and the remainder of the unperturbed energy $c - n$ distributed among the atoms.

For large r and/or c, the results[24] for the ground state and the first few excited states of the system are qualitatively of the form shown in Fig. A.1. We see that they look like harmonic oscillator eigenfunctions. Under these conditions, it is indeed possible to approximate the finite difference equation (A.4.12) by a differential equation in n which resembles the harmonic oscillator equation. The analogy then extends to the eigenfunctions (although there are differences due to different boundary conditions) of the ground

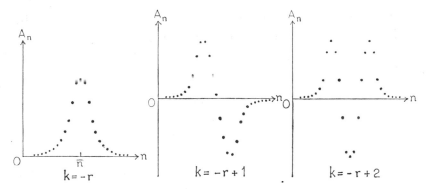

Figure A.1 Qualitative behavior of the amplitude $A_n^{(r,c,k)}$ to find n photons in the field in the stationary state $|r, c, k\rangle$ for large r and/or c, for the ground state and low-lying excited states [after M. Tavis and F. W. Cummings, *Phys. Rev.* **170**, 379 (1968)]

state and the lowest excited states (but *not* to highly excited states). The corresponding energy levels are approximately equally spaced.

If we compute the dispersion $\overline{n^2} - \bar{n}^2$ in the photon number in the ground state, we find (unless $c \gg r$, when almost all the energy is in the field) values of the order of $\bar{n}$, as would be the case for a Poisson distribution. Therefore, the field in the ground state resembles to some extent a coherent state [cf. (4.2.15) and Figs. A.1 and 7.6].

For $r = c \gg 1$, one finds that, in (A.4.14), $\varepsilon_{r,c-r} \approx -ar^{3/2}$, where a is a number of the order of unity. In particular, if $r = c = N/2$ (corresponding to an unperturbed state with all atoms in the upper level and no photon in the field), which is the case investigated by Scharf[25], the ground state energy is

$$E_{r,c,-r} \approx \frac{N}{2} - a\varkappa\left(\frac{N}{2}\right)^{3/2}. \qquad (A.4.15)$$

A first sight, this result seems to indicate that the system becomes unstable in the thermodynamic limit (volume $V \to \infty$, and $N \to \infty$, in such a way that the density N/V remains finite). However, one must remember, as was noted following (A.3.3), that the coupling constant $\varkappa$ in (A.3.4) is proportional to $V^{-\frac{1}{2}}$ (i.e., to $N^{-\frac{1}{2}}$). Thus the right side of (A.4.5) is proportional to N, and the system is stable.

On the other hand, for a sufficiently large value of the coupling constant, $\varkappa$, the last term in (A.4.15) might become dominant, depressing the ground state energy below that for the state with no photons in the field and all atoms in the lower level. This suggests the possibility of a *phase transition*. The behavior of the system in the thermodynamic limit and the existence of a phase transition were investigated by Hepp and Lieb[27], whose results will now be discussed.

A.5 THE THERMODYNAMIC LIMIT

In order to investigate the thermodynamic limit, we begin by rewriting the hamiltonian (A.3.4) in the form (indicating explicitly the dependence of various quantities on N)

$$H_N = a^+a + \frac{\varepsilon}{2} S_N^3 + \frac{\lambda}{2\sqrt{N}} (S_N^+ a + S_N^- a^+), \qquad (A.5.1)$$

where we have gone back to the Pauli spin vectors [cf. (8.2.9) and (8.3.4)],

$$\mathbf{S}_N = 2\mathbf{R}_N = \sum_{j=1}^{N} \boldsymbol{\sigma}(j), \qquad (A.5.2)$$

and the N-dependence of the coupling constant ($\varkappa = \lambda/\sqrt{N}$) has been rendered explicit. We have also allowed for deviations from resonance by introducing the parameter $\varepsilon = \omega_0/\omega$ [cf. (A.3.2)].

The procedure for analyzing the thermodynamic limit is similar to that applied by Thirring and Wehrl[28] to the BCS model of superconductivity. There is a considerable analogy between the two problems: as has been shown by Anderson[29], the BCS model can also be formulated in terms of spin operators.

The first problem in going over to the limit $N \to \infty$ is to define the limit of the state space (A.4.2) for the atoms,

$$\lim_{N \to \infty} \mathscr{H}_N = \mathscr{H}_\infty = \prod_{j=1}^{\infty} \otimes \mathbb{C}^2(j). \qquad (A.5.3)$$

The definition of the infinite tensor product of Hilbert spaces is due to von Neumann[30]; other applications to physics are described in references[31,32]. The space $\mathscr{H}_\infty$ is not separable, (i.e., it does not have denumerable dimension); it decomposes into a direct sum of orthogonal separable subspaces $\mathscr{H}_\infty(e)$,

$$\mathscr{H}_\infty = \sum_e \oplus \mathscr{H}_\infty(e). \qquad (A.5.4)$$

In order to define $\mathscr{H}_\infty(e)$, let us consider the eigenstates $|e\rangle$ of $\boldsymbol{\sigma} \cdot \hat{\mathbf{e}}$ for a single atom, where $\hat{\mathbf{e}}$ is a real (three-dimensional) unit vector:

$$(\boldsymbol{\sigma} \cdot \hat{\mathbf{e}}) |e\rangle = |e\rangle, \quad \langle e | e \rangle = 1. \qquad (A.5.5)$$

If (θ, φ) are the polar angles of $\hat{\mathbf{e}}$, the eigenstate $|e\rangle$ is explicitly given by

$$|e\rangle = e^{i\chi} \begin{pmatrix} \cos\dfrac{\theta}{2} \\ \sin\dfrac{\theta}{2} \, e^{i\varphi} \end{pmatrix}, \qquad (A.5.6)$$

where $e^{i\chi}$ is an arbitrary phase factor. In terms of the two-level atom, $|e\rangle$ represents a coherent superposition of the upper and lower states. We have

$$\langle e' | e \rangle = e^{i\psi} \sqrt{\frac{1 + \hat{\mathbf{e}} \cdot \hat{\mathbf{e}}'}{2}}, \tag{A.5.7}$$

where ψ is well defined once the phase factors associated with $|e\rangle$ and $|e'\rangle$ [cf. (A.5.6)] are given (e.g., $\psi = 0$ for $|e\rangle = |e'\rangle$).

Note that (A.5.5) is a particular case of (A.2.15) corresponding to spin $\frac{1}{2}$; also, (A.5.7) is a particular case of (A.2.20) and (A.2.21) with $r = \frac{1}{2}$.

Let us now consider an N-atom state, of the form

$$|\{e\}\rangle = \prod_j \otimes |e_j\rangle, \tag{A.5.8}$$

where $\{e\}$ stands for a sequence $e_1, e_2, ..., e_j, ...$ (one for each atom). For finite N, we have

$$\langle \{e'\} | \{e\}\rangle = \prod_j \langle e'_j | e_j \rangle, \tag{A.5.9}$$

where each scalar product is given by (A.5.7). What happens as $N \to \infty$? There are three possibilities:

(i) The infinite product $\prod_{j=1}^{\infty} \langle e_j | e'_j \rangle$ is absolutely convergent, i.e.,

$$\sum_{j=1}^{\infty} |1 - \langle e'_j | e_j \rangle| = \sum_{j=1}^{\infty} \left| 1 - e^{i\psi_j} \sqrt{\frac{1 + \hat{\mathbf{e}}_j \cdot \hat{\mathbf{e}}'_j}{2}} \right| < \infty, \tag{A.5.10}$$

so that $\hat{\mathbf{e}}_j$ and $\hat{\mathbf{e}}'_j$ are asymptotically parallel, and $\psi_j \to 0$. In this case, we say that $|\{e\}\rangle$ and $|\{e'\}\rangle$ are *equivalent*,

$$|\{e'\}\rangle \approx |\{e\}\rangle, \tag{A.5.11}$$

and they are said to belong to the same *equivalence class* $C\{e\}$.

Two vectors $|\{e\}\rangle$ and $|\{e'\}\rangle$ that differ only by a finite number of components $|e_j\rangle$ are obviously equivalent. By linear combinations of such vectors and by completion, we generate a *separable* Hilbert space $\mathcal{H}_\infty(e)$. This Hilbert space contains exactly those vectors that are equivalent to $|\{e\}\rangle$. The vectors obtained from $|\{e\}\rangle$ by "flipping" a finite number of spins are dense in $\mathcal{H}_\infty(e)$.

(ii) The infinite product is not absolutely convergent, but becomes so by removing the phase factors $e^{i\psi_j}$,

$$\sum_{j=1}^{\infty} \left| |1 - |\langle e'_j | e_j \rangle|| \right| = \sum_{j=1}^{\infty} \left| 1 - \sqrt{\frac{1 + \hat{\mathbf{e}}_j \cdot \hat{\mathbf{e}}'_j}{2}} \right| < \infty. \tag{A.5.12}$$

We then say that $|\{e\}\rangle$ and $|\{e'\}\rangle$ are *weakly equivalent*,

$$|\{e'\}\rangle \underset{W}{\approx} |\{e\}\rangle, \qquad (A.5.13)$$

and all such $|\{e'\}\rangle$ define the *weak equivalence class* $C_W\{e\}$. An example in which (A.5.13) is valid, although (A.5.11) is not, is obtained by multiplying each $|e_j\rangle$ in (A.5.8) by a phase factor $e^{i\lambda_j}$, such that $\sum_{j=1}^{\infty} \lambda_j \to \infty$; the resulting vector $|\{e'\}\rangle$ differs from $|\{e\}\rangle$ by an "infinite phase factor". In this case, (A.5.9) is meaningless, and the only consistent interpretation is to *define*

$$\langle\{e'\} \mid \{e\}\rangle = 0. \qquad (A.5.14)$$

Thus, although $\mathcal{H}_\infty(e)$ and $\mathcal{H}_\infty(e')$ may be physically equivalent, they are orthogonal Hilbert spaces.

(iii) Finally, if even (A.5.12) diverges, we again set

$$\langle\{e'\} \mid \{e\}\rangle = 0, \qquad (A.5.15)$$

and the vectors are not even weakly equivalent.

The decomposition (A.5.4) shows that $\mathcal{H}_\infty$ is a direct sum of separable orthogonal Hilbert spaces $\mathcal{H}_\infty(e)$. Since there are uncountably many equivalence classes, $\mathcal{H}_\infty$ is nonseparable.

Let us now apply these results to the thermodynamic limit for our system. The idea is that, by going to the limit along different "directions" $\mathcal{H}_\infty(e)$, we may get physically different results.

Let us take as our first choice

$$|\{e\}\rangle = |\{e'\}\rangle, \quad \hat{\mathbf{e}}'_j = (0, 0, -1), \quad |e'_j\rangle = \begin{pmatrix} 0 \\ 1 \end{pmatrix}, \quad \text{all } j, \quad (A.5.16)$$

corresponding to "all spins down", i.e., all atoms in the lower level. The corresponding space $\mathcal{H}_\infty(e')$ is generated by switching finitely many "spins" (atom excitations) and taking the closure. It can then be shown[27] that, in this Hilbert space, the thermodynamic limit of the hamiltonian (A.4.16) yields

$$H_N + \frac{\varepsilon}{2} N \xrightarrow[\substack{W \text{ in } \mathcal{H}_\infty(e')}]{N \to \infty} H'_B = a^+ a + \varepsilon N(e'), \qquad (A.5.17)$$

where "W" means *weak limit* and the "Bogoliubov-type" effective hamiltonian H'_B describes "quasi-particle" excitations, with

$$N(e) = \tfrac{1}{2} \sum_j [1 - \boldsymbol{\sigma}(j) \cdot \hat{\mathbf{e}}_j], \qquad (A.5.18)$$

corresponding to the *number of atomic excitations* ("spin-flips").

By comparing (A.5.17) and (A.5.18) with (A.5.1), we see that it amounts to the statement that

$$\frac{1}{\sqrt{N}} \sum_{j=1}^{N} \sigma_1(j) \frac{1}{W \text{ in } \mathscr{H}_\infty(e')} \to 0 \quad \text{ao } N \to \omega, \quad \text{(A.5.19)}$$

due to the factor $1/\sqrt{N}$. This means that we get a "trivial" limit, in which the interaction effectively vanishes. By (A.5.17), the ground-state energy per particle in the thermodynamic limit is

$$\mathscr{E}_0'(\lambda, \varepsilon) = -\frac{\varepsilon}{2}, \quad \text{(A.5.20)}$$

corresponding to the energy in the lower atomic level; also, the quasi-spin excitations have the energy ε.

The results given in § A.4 suggest that, for sufficiently large coupling constant λ, we may get a different ground state by going to the thermodynamic limit along a different "direction" $\mathscr{H}_\infty(e'')$, corresponding to some coherent superposition state for the atoms and a large photon excitation, with non-vanishing expectation value for the field.

We therefore make a "Bogoliubov-type" transformation of the photon operators, taking

$$a = b_N - \alpha \sqrt{N}, \quad \text{(A.5.21)}$$

where α is a (generally complex) c-number, to be determined later; thus

$$[b_N, b_N^+] = [a, a^+] = 1. \quad \text{(A.5.22)}$$

Note that the vacuum state for the b_N operators, $b_N||0\rangle\rangle_{b_N} = 0$, would correspond to a coherent state [cf. (3.1.10)] for a, with an average number of photons $= O(N)$, i.e., a finite density in the thermodynamic limit, which explains the above "Ansatz".

Substituting (A.5.21) in (A.5.1), we find that the result may be written as follows:

$$H_N = H_{N,B}'' + N\left(|\alpha|^2 - \frac{\gamma}{2}\right) + V_N, \quad \text{(A.5.23)}$$

where

$$H_{N,B}'' = b_N^+ b_N + \frac{\gamma}{2} \sum_{j=1}^{N} [1 - \sigma(j) \cdot \hat{e}''], \quad \text{(A.5.24)}$$

$$V_N = \frac{\lambda}{2\sqrt{N}} \left[\left(S_N^+ - \frac{2N\alpha^*}{\lambda}\right) b_N + \left(S_N^- - \frac{2N\alpha}{\lambda}\right) b_N^+\right], \quad \text{(A.5.25)}$$

15*

and γ and the unit vector $\hat{e}''$ are such that

$$\gamma\hat{e}'' = [\lambda(\alpha + \alpha^*), i\lambda(\alpha - \alpha^*), -\varepsilon], \qquad (A.5.26)$$

$$(\hat{e}'')^2 = 1. \qquad (A.5.27)$$

In the thermodynamic limit, we want the "residual interaction" term V_N to converge weakly to zero [cf. (A.5.19)] in the space $\mathcal{H}_\infty(e'')$ that will be defined by the unit vector $\hat{e}''$ (to be determined below), i.e., we must have

$$\langle\{e''\}|\, V_N\, |\{e''\}\rangle \to 0 \quad \text{as} \quad N \to \infty. \qquad (A.5.28)$$

Since [cf. (A.5.6), (A.5.8), (A.2.11)], for finite N,

$$\langle\{e''\}|\, S_N^\pm\, |\{e''\}\rangle = N\langle e''|\, \sigma^\pm\, |e''\rangle = Ne'_\pm, \qquad (A.5.29)$$

where

$$e''_\pm = e''_1 \pm ie''_2, \qquad (A.5.30)$$

it follows from (A.5.25) and (A.5.28) that we must have

$$e''_+ = 2\alpha^*/\lambda, \quad e''_- = 2\alpha/\lambda; \quad e''_+ e''_- = 4\,|\alpha|^2/\lambda^2 = 1 - (e''_3)^2. \qquad (A.5.31)$$

Comparing this with (A.5.26), we get

$$\gamma = \lambda^2. \qquad (A.5.32)$$

Substituting these results in (A.5.26) and (A.5.27), we find the analogue of the gap equation of the BCS theory,

$$\lambda^2 = \sqrt{\varepsilon^2 + \lambda^4[1 - (e''_3)^2]}. \qquad (A.5.33)$$

In order for this to determine a real direction, with $|e''_3| \le 1$, we must have

$$\lambda^2 \geqq \varepsilon, \qquad (A.5.34)$$

which gives the required strength of the coupling constant ($\varepsilon = 1$ at resonance).

If (A.5.34) is satisfied, we can find e''_3 from (A.5.33), and (A.5.31) then yields

$$|\alpha| = \frac{\sqrt{\lambda^4 - \varepsilon^2}}{2\lambda}. \qquad (A.5.35)$$

Substituting (A.5.28), (A.5.32) and (A.5.35) in (A.5.23) and (A.5.24), we find, in the place of (A.5.17),

$$H_N - N\mathscr{E}''_o \xrightarrow[W \text{ in } \mathcal{H}_\infty(e'')]{N \to \infty} H''_B = b^+b + \lambda^2 N(e'') \qquad (A.5.36)$$

where

$$\mathscr{E}''_o(\lambda, \varepsilon) = -\frac{\varepsilon}{2}\left(\frac{\lambda^2}{2\varepsilon} + \frac{\varepsilon}{2\lambda^2}\right), \quad \lambda^2 \geqq \varepsilon. \qquad (A.5.37)$$

and $N(e'')$ is given by (A.5.18), with $\hat{e}_j = \hat{e}''$.

The right side of (A.5.36) is a new Bogoliubov effective hamiltonian H_B'', with b^+b representing the number of "quasi-photon" excitations, and $N(e'')$ the number of "quasi-spin" atomic excitations, with excitation energy λ^2 greater than the corresponding excitation energy ε for H_B' in (A.5.17). Since the expression within brackets in (A.5.37) is $= 1$ for $\lambda^2 = \varepsilon$ and is >1 for $\lambda^2 > \varepsilon$, we have $|\mathscr{E}_0''| > |\mathscr{E}_0'|$ [cf. (A.5.20)] for $\lambda^2 \geqq \varepsilon$, so that we indeed have a new ground state, with a "frozen-in" macroscopic electromagnetic field and nonzero average excitation energy for the atoms. This new ground state will be called "superradiant", although this term is employed here in a somewhat different sense[32a] from that of § 8.4 (it may be compared with the term "superconducting" state). The ground-state energy per particle $\mathscr{E}_0(\lambda, \varepsilon)$ is given by (A.5.20) in the "nonradiant" state ($\lambda^2 < \varepsilon$) and by (A.5.37) in the "superradiant" state ($\lambda^2 > \varepsilon$), as illustrated in Fig. A.2.

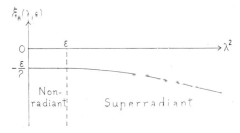

Figure A.2 Qualitative behavior of the ground-state energy per particle as a function of the coupling constant, illustrating the superradiant phase transition at $\lambda^2 = \varepsilon$.

Note that the projection of e'' on the (1, 2) plane is determined only in modulus, but not in direction, by (A.5.31) and (A.5.35), so that we actually have a continuous one-parameter family of solutions,

$$e_+''(\varphi) = \sqrt{1 - \varepsilon^2/\lambda^4}\, e^{i\varphi}, \qquad 0 \leqq \varphi < 2\pi. \tag{A.5.38}$$

This degeneracy is related with a broken symmetry, as happens in the BCS and other models: in the present case, broken rotation symmetry in the (1,2) plane. Correspondingly, we should use the notations $\mathscr{H}_\infty(e_\varphi'')$, $H_B''(\varphi)$. Different solutions (A.5.38) correspond to vectors that are not weakly equivalent [cf. (A.5.15)], and therefore[31] to unitarily inequivalent representations of the spin algebra.

Precise estimates for the eigenvalues and the errors for large N may be obtained[27]. The ground-state energy per particle corresponds to the lowest eigenvalue of the tridiagonal matrix discussed in § A.4. A theorem due to Hepp and Lieb[27] allows one to obtain asymptotic estimates for the lowest eigenvalue, in agreement with the above results.

The thermodynamic properties of the system can also by investigated, by evaluating the free energy per particle

$$f_N(\beta, \varepsilon, \lambda) = -\frac{1}{\beta N} \ln \text{Tr} \, (e^{-\beta H_N}), \quad \beta = \frac{1}{k_B T}, \quad \text{(A.5.39)}$$

in the thermodynamic limit. This can be done in the Dicke basis, approximating H_N by the lowest eigenvalue. The result[27] is that f_N approaches two different analytic functions, f' or f'', depending on whether $T \gtreqless T_c$ or $T < T_c$, where the *critical temperature* T_c is defined by

$$\beta_c = \frac{1}{k_B T_c} = \frac{2}{\varepsilon} \tanh^{-1}\left(\frac{\varepsilon}{\lambda^2}\right), \quad \lambda^2 > \varepsilon. \quad \text{(A.5.40)}$$

The specific heat is discontinuous at T_c, so that we have indeed a *second-order phase transition*.

Thermal Green's functions may also be computed[27] from the Bogoliubov effective hamiltonians; for other variables, the spin-wave approach of Wehrl[33] leads to better results.

The existence of a phase transition in the present model is connected with the very long range of the interaction: all atoms see essentially the same field, leading to very strong correlations. There is a considerable degree of analogy between the "superradiant" phase transition and the superconducting phase transition in the BCS model, which in its turn is known to be analogous to the ferromagnetic or Van Der Waals gas phase transitions in the mean field approximation.[34] The lowering of the ground-state energy in the "superradiant" ground state results from the interaction energy $-\mathbf{P} \cdot \mathbf{E}$; we have a macroscopic effective dipole moment, and the direction $\mathbf{e}''$ in (A.5.31) is determined by the alignment of this dipole moment with the "local field". An experimental verification seems to be difficult, however, because in practice one usually has $\lambda^2 \ll \varepsilon$.

It can be asked whether a similar phase transition occurs in a laser at its threshold of operation (corresponding to the critical density of population inversion). As remarked by Graham and Haken[35] and by De Giorgio and Scully[36], there exists indeed a considerable analogy between the existing versions of the quantum theory of the laser and the macroscopic Landau theory of phase transitions, which corresponds to a mean field theory. This is not surprising, since the usual laser theories employ the self-consistent field approximation. The connection between these results and non-equilibrium thermodynamics has also been investigated by Graham and Haken.[37] Hepp and Lieb[37a] have recently extended their treatment to the laser problem by coupling Dicke's model system to pumping and loss reservoirs;

this does lead to a phase transition at the laser threshold. Their model also allowed them to derive quantum Langevin equations of the type discussed in § 7.7(b).

A.6   DYNAMICS OF SUPERRADIANT EMISSION

In §§ A.4 and A.5, we considered only stationary states of the atom-field system. Dicke's original discussion of superradiance was in terms of the emission of a pulse by the system [cf. § 8.4(b)]. Let us now discuss improved treatments for the non-stationary case, going beyond the semiclassical approximation and first-order perturbation theory.

Bonifacio and Preparata[38] have treated the evolution of nonstationary states for the system described by the hamiltonian (A.3.4), i.e., a set of atoms placed in a cavity, interacting with a single resonant mode, and occupying a region of size much smaller that the wavelength. In contrast with Dicke's spontaneous emission problem, the energy is not radiated into free space, but rather it is conserved within the cavity, so that we expect it to oscillate back and forth between the field and the atomic system.

We consider an initial (nonstationary) state of the form $||0\rangle\rangle\,|r, m\rangle$, with no photons and the atomic system in the Dicke state $|r, m\rangle$. Two cases are of special interest:

(i) *Fully excited* initial state: $|r, m\rangle = |\dfrac{N}{2}, \dfrac{N}{2}\rangle$,

(ii) *Superradiant* initial state: $|r, m\rangle = |\dfrac{N}{2}, 0\rangle$,

where N is assumed to be very large (of course, values of r and m sufficiently close to these values will do as well).

Since the conserved quantum number c equals m for the above states [cf. (A.4.8)], they will evolve into a time-dependent superposition of states $||n\rangle\rangle\,|r, m - n\rangle$. The probability amplitude for finding n photons in the field at time t is

$$p(n, t) = \langle\langle n||\,\langle r, m - n|\,e^{-iHt}\,|r, m\rangle\,||0\rangle\rangle. \tag{A.6.1}$$

The average number of photons $\bar{n}(t)$ at time t gives the field intensity, and one is also interested in the photon statistics.

For very short times, the solution is given by first-order perturbation theory, which gives Dicke's radiation rate (8.4.2):

$$p(n, t) = -i\delta_{n, 1}\,\sqrt{r(r + 1) - m(m - 1)}\,t, \tag{A.6.2}$$

so that

$$\bar{n}(t) = (r^2 - m^2 + r + m)\,t^2. \tag{A.6.3}$$

The classical-limit approximation (8.4.9) that led to (8.4.17) corresponds to approximating the intensity by [cf. (8.4.11)]

$$I \approx I_0(r^2 - m^2) = I_0 r^2 \sin^2 \theta, \tag{A.6.4}$$

i.e., to neglecting the term $(r + m)$ in (A.6.3) in comparison with $r^2 - m^2$. For a superradiant initial state $(r \gg m)$, this is clearly a good approximation; however, for a fully excited initial state $(r = m)$, it is obviously incorrect.

Indeed, the fully excited initial state corresponds to $\theta = 0$ in (8.4.9), and this is an equilibrium state [corresponding to the unstable vertical pendulum position in (8.4.15)], with vanishing radiation rate (A.6.4) in the classical approximation. The missing terms $r + m$ in (A.6.4) represent the quantum spontaneous emission, which "sets the pendulum into motion", no longer allowing an unstable equilibrium position. As soon as the pendulum falls down enough so that $r - m \gg 1$, the other term $r^2 - m^2$ in (A.6.3), which corresponds to the classical polarization, will again dominate.

Thus, we can think of (A.6.3) as representing the combined effect of two sources for the field: (a) the classical macroscopic polarization source $r^2 - m^2$; (b) the quantum spontaneous emission source $r + m$. The different nature of these two sources is reflected in the difference between the short-time photon statistics corresponding to the fully excited and superradiant initial states.

For short times, one finds[38], in the superradiant case,

$$\overline{(\Delta n^2)} \approx \bar{n}, \tag{A.6.5}$$

i.e., the Poisson statistics characteristic of a coherent state [cf. (4.2.15)], whereas, for the fully excited case,

$$\overline{(\Delta n)^2} \approx \bar{n}(\bar{n} + 1), \tag{A.6.6}$$

i.e. we find the Bose statistics characteristic of a chaotic field [cf. (4.2.13)]. In the superradiant case, the initial state gives rise to a macroscopic transverse dipole, which, as a classical source, drives the field into a coherent state; in the fully excited state, the field is driven by chaotic incoherent radiation due to spontaneous emission (cf., also, the discussion at the end of § A.2).

These results are confirmed by the approximate solutions[38] obtained without the restriction to short times, in which $\bar{n}(t)$ is found to be given by Jacobian elliptic functions. The periodic character of these solutions reflects the energy oscillation between the atoms and the cavity field. Actually, the exact solutions are not periodic, but only quasi-periodic[39], corresponding to the fact that the energy levels for the stationary solutions (§ A.4) are not exactly equally spaced.

For the fully excited initial state, the approximate solution represents a periodic train of pulses, each of which coincides in shape with the classical hyperbolic secant solution (8.4.17), with a width $\propto r^{-\frac{1}{2}}$. However, instead of an infinite build-up time, the separation between consecutive pulses is of the order $\ln r$ times larger than the width of a single pulse. For the superradiant initial state, the separation between pulses is of the same order as the width of each pulse, so that the pulse train resembles the square of a sine function.

The above discussion is still restricted to a cavity with dimensions much smaller than the wavelength. The effects of removing this restriction have already been discussed in § 8.4(c). We saw there that superradiant states can still be defined, but now for radiation in a given direction $\mathbf{k}$; emission of radiation in directions $\mathbf{k}' \neq \mathbf{k}$ tends to produce transitions into states with lower cooperation number. As shown by Dicke[40], for excitation by an incident plane wave in the direction $\mathbf{k}$ and observation in the direction $\mathbf{k}'$, the small-cavity superradiant enhancement factor $N/4$ per atom [cf. (8.4.4)] has to be weighted with another factor, given by

$$|\langle e^{i(\mathbf{k}-\mathbf{k}')\cdot\mathbf{x}}\rangle|^2, \tag{A.6.7}$$

where the angular brackets denote an average over atomic positions $\mathbf{x}$. This factor represents the directivity of a set of classical dipole oscillators excited by a plane wave, and it is strongly dependent on the geometry.

For a rod-shaped cavity with cross-sectional area A, and $\mathbf{k}$ along the axis, the directivity factor confines most of the radiation into a small solid angle[41]

$$\Delta\Omega \approx 4\pi\lambda^2/A \tag{A.6.8}$$

around the axis, where λ is the wavelength ("end-fire modes"). Within this solid angle, we get the full enhancement factor (8.4.4) for superradiant emission, i.e., $N/4$ per atom. Since the spontaneous emission rate per atom into $\Delta\Omega$ is $\gamma\,\Delta\Omega/4\pi$, where γ is the total rate for a single atom [cf. (8.4.21)], the actual Dicke enhanced radiation rate for superradiant emission per atom becomes

$$\gamma_D = \tfrac{1}{4}N\gamma\lambda^2/A. \tag{A.6.9}$$

It was pointed out by Arecchi and Courtens[42] that (A.6.9) still contains implicit assumptions that make it unrealistic for very large N: they showed that there exists a *maximum cooperation number* N_c and a *maximum cooperation length* l_c. To see this, we note that $\tau_D = 1/\gamma_D$, which represents the lifetime for superradiant decay per atom, must be much larger than (a) the optical period $2\pi/\omega$; (b) the travel time l/c of the emitted radiation over the sample length l. Condition (a) arises from the fact that the first-order perturbation theory result is obtained by averaging over many optical periods. Condition (b) arises because the perturbation calculation treats the emitted radiation as a plane wave, neglecting damping over the distance l.

At optical frequencies, condition (b) is the more stringent one: in order for the atoms in a length l of the sample to cooperate, each of them must be exposed to the radiation from all the others before the decay process is completed. To find the actual decay rate γ_c for a sample of length $l > c\tau_D$, we can apply a self-consistency argument.

Let ϱ be the density of two-level atoms. The number of atoms covered by the radiation from one of them during the actual decay time $\tau_c = 1/\gamma_c$ is

$$N_c = c\tau_c A\varrho. \tag{A.6.10}$$

The Dicke radiation rate per atom (A.6.9) applies only to these atoms, so that the actual rate γ_c per atom is

$$\gamma_c = \frac{1}{\tau_c} = \frac{1}{4} N_c \gamma \lambda^2 / A,$$

which, together with (A.6.10), yields

$$\gamma_c = \frac{1}{\tau_c} = \frac{1}{2} \sqrt{c\varrho\gamma\lambda^2}, \tag{A.6.11}$$

and, substituting in (A.6.10),

$$N_c = \frac{2A}{\lambda} \sqrt{\frac{c\varrho}{\gamma}}. \tag{A.6.12}$$

The number N_c is the *maximum cooperation number*, i.e., the maximum number of atoms that can cooperate to give superradiant emission; if a sample has $N > N_c$, all N atoms may emit, but only N_c can cooperate. The time τ_c is the *cooperation time*, and the *maximum cooperation length* is

$$l_c = c\tau_c. \tag{A.6.13}$$

This is also the maximum sample length that can be excited into a superradiant state by an incident pulse. For the 6943 Å transition in ruby with 0.05% concentration of Cr ions, one finds $l_c \approx 1$ mm.

A more detailed analysis of superradiant emission without the restriction to a small cavity has been carried out by Bonifacio, Schwendimann and Haake[43]. By choosing as a sample a thin rod of length $l \ll l_c$ (but l can still be $\gg \lambda$), they managed to retain the strong coupling only to a single axial ("end-fire") mode. In contrast with the laser, a low-Q cavity is taken, so that the photons escape very fast, giving rise to a single emitted pulse. The system is described by a density operator, which initially represents the field in the vacuum state and not correlated with the atoms. These assumptions lead to a master equation, which is numerically solved for a superradiant initial state, as well as for the fully excited case.

For a superradiant initial state, the results are very close to the semi-classical ones: the shape of the emitted pulse is indistinguishable from the hyperbolic secant solution within the accuracy of the calculation. Within the high-intensity region of the pulse, the field may be considered to be in a coherent state. The fluctuations in the emitted intensity and in the energy of the atoms are small.

For a fully excited initial state, however, the results are very different from the semiclassical ones. The shape of the emitted pulse is still similar to a hyperbolic secant, but the peak intensity is reduced by about 23 % as compared with the semiclassical prediction. The intensity fluctuation is much larger than that for a coherent state. The dispersion $\langle R_3^2 \rangle - \langle R_3 \rangle^2$ of the total atomic energy takes values of the order N^2 near the peak of the pulse. There are large atom-atom and atom-field correlations at all times. These large quantum fluctuations may be regarded as a kind of amplified spontaneous emission noise arising from the incoherent emission in the initial state. They persist as long as the initial occupancy of the upper level differs from N at most by a number of order unity. This corresponds to the stochastic character associated with the spontaneous emission of the first few photons. These results have been confirmed by analytical calculations[44].

A general discussion of the domain of validity of classical approximations for a set of two-level atoms interacting with the radiation field (both for free decay and for forced oscillations) has recently been given by Senitzky[45].

The behavior of the emission rate and the angular distribution of the emitted radiation for large cavities of various shapes, such that the coupling is not restricted to a single mode of the field, have been investigated by Rehler and Eberly[46]. Their basic assumption is that the atomic state may be represented not just initially, but at all times, by a direct product of the type (A.2.38). According to the above discussion, this approach may be applied to a superradiant initial state, but not to a fully or almost fully excited one.

References

1. See the reviews by C. H. Townes, in *Contemporary Physics*, vol. I, International Atomic Energy Agency, Vienna (1969), and N. Bloembergen, in *Quantum Optics*, eds. S. M. Kay and A. Maitland, Academic Press, New York (1970).
2. See the review article by A. J. De Maria, in *Progress in Optics*, ed. E. Wolf, vol. IX, North-Holland Publishing Co., Amsterdam (1971).
3. R. E. Slusher and H. M. Gibbs, *Phys. Rev. A* **5**, 1634 (1972).
4. H. M. Gibbs and R. E. Slusher, *Phys. Rev. A* **6**, 2326 (1972).
5. H. C. Torrey, *Phys. Rev.* **76**, 1059 (1949).

16*

6. G. B. Hocker and C. L. Tang, *Phys. Rev. Letters* **21**, 591 (1968); *Phys. Rev.* **184**, 356 (1969).

7. R. G. Brewer and R. L. Shoemaker, *Phys. Rev. Letters* **27**, 631 (1971).

8. E. L. Hahn, *Phys. Rev.* **80**, 580 (1950); G. E. Pake, in *Solid State Physics*, F. Seitz and D. Turnbull, eds., Vol. 2, Academic Press, New York (1956), p. 32.

9. R. G. Brewer and R. L. Shoemaker, *Phys. Rev. A* **6**, 2001 (1972).

10. R. L. Shoemaker and R. G. Brewer, *Phys. Rev. Letters* **28**, 1430 (1972).

11. See also the extensive review article by S. Stenholm, *Physics Reports* **6**, 1 (1973).

12. J. M. Radcliffe, *J. Phys. A* **4**, 313 (1971).

13. F. T. Arecchi, E. Courtens, R. Gilmore and H. Thomas, *Phys. Rev. A* **6**, 2211 (1972).

14. K. Gottfried, *Quantum Mechanics*, vol. I, W. A. Benjamin, Inc., New York (1966), p. 213.

15. See, e.g., E. Hille, *Analytic Function Theory*, Vol. 1, Ginn & Co., Boston, Mass. (1959), p. 38.

16. E. Inönü and E. P. Wigner, *Proc. Nat. Acad. Sci. U.S.* **39**, 510 (1953); E. J. Saletan, *J. Math. Phys.* **2**, 1 (1961); E. Inönü, in *Group Theoretical Concepts and Methods in Elementary Particle Physics*, ed. by F. Gürsey, Gordon and Breach, N. Y. (1964), p. 391.

17. I. R. Senitzky, *Phys. Rev.* **111**, 3 (1958).

18. N. Bohr, *Nature* **121**, 580 (1928); W. Heisenberg, *The Physical Principles of the Quantum Theory*, Univ. of Chicago Press, Chicago, Ill. (1930), p. 46.

19. I. R. Senitzky, *loc. cit.*; R. Bonifacio, P. Schwendimann, and F. Haake, *Phys. Rev. A* **4**, 854 (1971).

20. R. Bonifacio and G. Preparata, *Phys. Rev. A* **2**, 336 (1970).

21. J. Schwinger, in *Quantum Theory of Angular Momentum*, eds. L. C. Biedenharn and H. Van Dam, Academic Press, N.Y. (1965).

22. N. Bloembergen, *Nonlinear Optics*, W. A. Benjamin, N.Y. (1965).

23. V. Ernst and P. Stehle, *Phys. Rev.* **176**, 1456 (1968); cf. also J. H. Eberly, *Am. J. Phys.* **40**, 1374 (1972).

24. M. Tavis and F. W. Cummings, *Phys. Rev.* **170**, 379 (1968); **188**, 692 (1969).

25. G. Scharf, *Helv. Phys. Acta* **43**, 806 (1970).

26. W. R. Mallory, *Phys. Rev.* **188**, 1976 (1969).

27. K. Hepp and E. H. Lieb, *Ann. Phys.* (*N.Y.*) **76**, 360 (1973); cf. also Y. K. Wang and F. T. Hioe, *Phys. Rev, A*, 7, 831 (1973).

28. W. Thirring and A. Wehrl, *Commun. Math. Phys.* **4**, 303 (1967).

29. P. W. Anderson, *Phys. Rev.* **112**, 1900 (1958).

30. J. von Neumann, *Compositio Math.* **6**, 1 (1939).

31. M. C. Reed, in *Brandeis Lect. on Elem. Part. Phys. and Quantum Field Theory*, eds. S. Deser, M. Grisaru, and H. Pendleton, Vol. 2, M.I.T. Press, Cambridge, Mass. (1970).

32. T. W. B. Kibble, *J. Math. Phys.* **9**, 315 (1968).

32(a). The atomic system approaches a coherent atomic state $|\theta, \varphi> = |\hat{\mathbf{e}}''>$ with $r = N/2$, so that its radiation rate (A 2.32) is $O(N^2)$; however, θ only approaches $\pi/2$ for $\lambda^2 \gg \varepsilon$.

33. A. Wehrl, *Commun. Math. Phys.* **23**, 319 (1971).

34. Cf. e.g., H. E. Stanley, *Introduction to Phase Transitions and Critical Phenomena*, Oxford Univ. Press, N.Y. (1971).

35. R. Graham and H. Haken, *Z. Physik* **213**, 420 (1968); **237**, 31 (1970).

36. V. De Giorgio and M. O. Scully, *Phys. Rev. A* **2**, 1170 (1970).

37. R. Graham and H. Haken, *Z. Physik* **243**, 289 (1971); **245**, 141 (1971); R. Graham, in *Springer Tracts in Modern Physics*, vol. 65, Springer-Verlag, Berlin and New York (1972).

37 (a). K. Hepp and E. H. Lieb, to be published.

38. R. Bonifacio and G. Preparata, *Phys. Rev. A* **2**, 336 (1970).

39. D. F. Walls and R. Barakat, *Phys. Rev. A* **1**, 446 (1970); I. R. Senitzky, *Phys. Rev. A* **2**, 2046 (1970).

40. R. H. Dicke, *Phys. Rev.* **93**, 99 (1954).

41. Cf. e.g., I. D. Abella, N. A. Kurnit and S. R. Hartmann, *Phys. Rev.* **141**, 391 (1966), Appendix C.

42. F. T. Arecchi and E. Courtens, *Phys. Rev. A* **2**, 1730 (1970).

43. R. Bonifacio, P. Schwendimann and F. Haake, *Phys. Rev. A* **4**, 302, 854 (1971).

44. R. Bonifacio and M. Gronchi, *Lett. Nuovo Cimento* **1**, 1105 (1971); *Nuovo Cimento* **8, B**105 (1972); V. DeGiorgio and F. Ghielmetti, *Phys. Rev. A* **4**, 2415 (1971); F. Haake and R. J. Glauber, *Phys. Rev. A* **5**, 1457 (1972).

45. I. R. Senitzky, *Phys. Rev. A* **6**, 1175 (1972).

46. N. E. Rehler and J. H. Eberly, *Phys. Rev. A* **3**, 1735 (1971); J. H. Eberly, *Am. J. Phys.* **40**, 1374 (1972).

Index

Absorption 23, 94, 95
  cross-section 96
  length 200
Absorptive part 189, 194
Active atoms 90, 94, 174
  density of 97
Adiabatic approximation 138–139, 149,
    157, 166
Airy diffraction pattern 11
Amplifying medium 192, 201–202
Amplitude equation 105–107
Analytic functional 57
Analytic signal 4, 18, 65, 104
Angular correlation between photons 183
Angular momentum
  Schwinger's theory of 219
  vector model of 179, 218
Annihilation operators 18, 39, 41–42, 65,
    143, 218
  of coherent states 46
  relation with Q and P operators 43
Antinormally-ordered
  characteristic function 62, 63
  product 28, 161–162
Antiresonant terms 100, 118, 122, 146
Area of pulse 193, 197
Area theorem 196, 198, 208
Atom loss reservoirs 101, 108, 115
Atom, two-level, *see* Two-level atoms
Atomic coherence effects 171, 178, 207
Atomic line shape factor 117, 119, 188
Attenuating medium 192, 202
Axial mode 92, 103, 143

Backscattering 191
Baecklund transformation 203
Baker-Hausdorff lemma 45
BCS model of superconductivity 224, 228,
    230
Beer's law 198, 200
Biological competition 132, 204

Blackbody radiation 4, 30, 34, 76
  see also Thermal light
Bloch equations 115–116, 173, 179, 185,
    189
Bloch sphere 179, 186, 195, 212, 214
Bloch-type damping terms 190
Bochner-Schwartz theorem 33–34
Bogoliubov effective hamiltonian 226,
    229
Bogoliubov transformation 227
Boltzmann factor 96
Bose distribution 71, 78
Bose-Einstein condensation 39, 183
Bose-Einstein statistics 76, 81, 232
Bounded operator 53, 58, 61, 69
Brewster-angle windows 89, 100
Broadening
  collision 95–96, 138
  Doppler 93, 95–96, 115, 132
  homogeneous 95, 117, 134
  inhomogeneous 96, 132, 183, 185, 188,
    209
Brownian motion 160, 164, 166
Bunching effects 80–81

Causality 31, 201–202
Central limit theorem 73
Chaotic field 73, 78, 81, 137, 232
Characteristic function 58, 62
  antinormally-ordered 62, 63
  normally-ordered 62, 63, 66, 78, 84–87
Chirping 188, 208
Classical model of superradiance 179,
    218, 232
Coherence
  and noiselessness 37, 48, 218
  area 3, 12, 82, 187
  atomic 171, 178, 207
  length 3
  propagation of 10
  relation to superfluidity and supercon

239

Coherence
    ductivity   39, 205
    spatial   10, 31
    temporal   8, 160
    time   2, 78, 80, 82
        and spectral width   2, 9
        for laser light   3
        for thermal light   3
    volume   3, 39
Coherence functions
    analytic properties   32
    and hermiticity   29
    and local commutativity   31
    and positive definiteness   32–34
    and spectral condition   31–32
    classical   6, 13
    for chaotic light   78–79, 87
    for ideal laser light   87
    general properties   28–34, 84–87
    higher-order   13, 25–27, 28
    in N-photon state   29
    inequalities for   32–34
    invariance properties   29–30
    mutual   6, 10, 11, 82
    reconstruction problem   34, 84–87
    relation to Wightman functions   29
    second-order   6, 13, 21
Coherent atomic states
    and classical limit   217
    as minimum-uncertainty states   212
    as rotated angular-momentum states
        212
    definition   210, 212
    non-orthogonality of   210–213
    overcompleteness of   213
    produced by classical field   215
    radiation rate from   216
    representations in terms of   214
    superradiant   217
Coherent fields
    and factorization of coherence functions
        35, 37, 39
    density operators for   38–39
    in higher order   36–37, 74–75
    in second order   34–36
        and single mode occupancy   39
Coherent light   7
Coherent-state representation
    diagonal, *see* Diagonal coherent-state

    representation
    of operators   53
    of state vectors   50–52
Coherent states
    and classical limit   44, 48
    and forced harmonic oscillator   48–49
    as displaced harmonic oscillator states
        45
    as limits of coherent atomic states   215
    as minimum-uncertainty states   47–48
    average number of photons in   44
    definition   41–42
    expansion in Fock states   43
    for a single mode   42
    for any number of modes   64
    Hamiltonians that preserve   48–49
    non-orthogonality of   49
    overcompleteness of   49
    produced by classical current   74
    representation in terms of, *see* Coherent-
        state representation
    scalar product of   49
    wave function of   47
Coincidence counting   25, 27
Collisions of the second kind   91
Combination tones   122
Commutation rules   18, 39
Competition among modes   126–132, 204
Completeness relation   50
Completely inverted state, *see* State, fully
    excited
Complex degree of coherence, *see* Degree
    of coherence, complex
Conditional probability for photon counts
    80
Conservation of probability   112, 190
Cooperation
    length   233–234
    number   176–177, 183, 221
        maximum   233–234
    time   234
Correlations
    atom-atom   217–218, 235
    atom-field   235
Correspondence principle   44, 68
Counterrotating wave   205
Coupling constant   145, 218, 223, 228–230
Coupling of modes in laser   127, 130
Creation operators   18, 39, 143, 218

Creation operators
  of coherent states 46
  relation with Q and P operators 43
Critical inversion density, *see* Population inversion density, critical
Critical temperature 230
Cross-spectral density 34
Current, classical 73–74

Damping, quantum theory of 139, 141
Degeneracy
  of Dicke states 176
  of product states 175
  parameter 4, 82
Degree of coherence
  complex 7, 33, 34, 80
    modulus of 7, 8, 10, 34, 80
    phase of 7, 8, 10
  higher-order 36–38
  temporal 8
Density operator 19, 21, 29, 53, 61, 110, 138
  equation of motion for 111
  for chaotic field 71–73
  for coherent field 38–39, 74
  reduced 39, 139–141, 143, 146, 203
  equation of motion for 149
Detailed balance 94, 152
Detuning 125, 135, 166, 191, 194
Diagonal coherent-state representation 54 to 68, 220
  for a single mode 54, 61, 63, 161
  for any number of modes 65
  for thermal light 72
Diagonal equations of motion 116
Diagonal representation 20
Dicke states 176, 209–210, 215, 217–218
  for large system 182
  radiation rate from 177, 216
Diffraction losses 92, 101
Diffusion constant 160, 163
Diffusion term 163
Dipole approximation 22, 25, 100, 109, 172, 215
Directivity factor 183, 233
Dispersion relations 5, 10
Dispersive part 189
Displacement operator 46
Distribution

Schwartz 57
  tempered 57, 78
Domain of dependence 202
Doppler broadening, *see* Broadening, Doppler
Doppler effect 99, 117, 132, 185, 200
Doppler width 96
Drift term 163

Einstein's A and B coefficients 94
Electric dipole approximation, *see* Dipole approximation
Emission counter 27–28, 31
Emission rate, *see* Radiation rate
End-fire modes 183, 233, 234
Energy density 94, 96, 189, 196
Energy storage 188, 195–196, 202
Ensemble average
  and single trials 83–84, 160
  and time average 6
  classical 5, 13
  quantum 19
Equivalence class 225
Ergodic fields 6
Extinction coefficient 106–107, 198

Factorization property 35, 37, 39
Feynman histories 84
Field loss reservoirs 101, 102, 127, 143, 167
Fluctuation-dissipation relations 168
Fluctuations
  in photon number 76–77
  intensity 76, 235
Fock representation
  of operators 38, 53, 143, 149
  of state vectors 50, 220
Fock states 18, 42–44, 50, 77, 86, 144, 210, 215
Fokker-Planck equation 163–168
Free energy 230
Free induction decay 186, 208–209
Free oscillations 106
Frequency conversion 219
Frequency equation 105–107, 132
Frequency modulation 188, 208
Frequency pulling 107, 120, 132, 188
Fully excited state, *see* State, fully excited
Functional derivative 78

Gain
  effective   129, 164
  overall   96, 127, 153
Gaussian distribution   72–73, 164–165
  moments of   78
Gaussian wave packet   47
Generalized phase-space representation   68
Ground state of Dickes' model   223, 227, 229
Group contraction   209, 214–215

Hamiltonian
  effective   226, 229
  for a two-level atom interacting with the field   109, 172
  for Dicke's model   174
  for photoelectric effect   22
  for quantized Dicke model   219
  for quantum theory of the laser   145
Hanbury Brown and Twiss
  effect   12–13, 81–82, 84, 171
  stellar interferometer   12
Heisenberg picture, see Picture, Heisenberg
Hermiticity   29
Hilbert-Schmidt
  norm   59, 61
  operator   59, 61
Hilbert spaces
  infinite tensor product of   224–226
  of analytic functions   53
  separable   64, 224, 225
Hilbert transforms   5
Hole burning   133–135
Homogeneous broadening   95, 117, 134
  lifetime   171, 190
Hyperbolic secant
  superradiant pulse   180–181, 233, 235
  pulse, 2π, see Pulse, 2π hyperbolic secant
Hysteresis   130

Incoherence, spatial   10
Incoherent light   7
Induced emission   94–95, 109, 122, 127, 137, 153, 164, 188
  see also Stimulated emission
Induced transition rate   122–123, 133, 159

Infinite tensor product of Hilbert spaces   224–226
Infrared catastrophe   69, 204
Inhomogeneous broadening   96, 132, 183, 185, 188, 209
  lifetime   171, 188, 190, 200, 208
In-phase component   105
In-quadrature component   105
Intensity
  average   5, 21
  correlations   12, 13, 28, 81, 84
  fluctuations   76
  instantaneous   5
Interaction picture, see Picture, interaction
Interference
  of independent beams   82–84
  spectroscopy   8–10
Interferometer
  Michelson   2, 8–9, 82
  Michelson's stellar   12
  Hanbury Brown and Twiss   12
Inversion, see Population inversion
Isometric operator   58

Josephson junctions   39

Korteweg-de Vries equation   203
Krylov-Bogoliubov method   104–105

L^2 norm   61
Lamb dip   135
Langevin equation   166, 168
Langevin forces   166–168
Laser
  amplitude of steady-state oscillation   98, 103, 107, 123, 125
  cavity   91
    free oscillations   106
  CW (continuous wave)   89
  frequency of oscillation   103, 121
  gas   89
  He–Ne   89, 90
  light
    coherence time   3
    degeneracy parameter   4
    ideal   73–75, 77, 80–81, 87, 144, 155
    intrinsic linewidth   99
  phase transition   204, 230
  ruby   89

Laser
 theory, quantum
  basic assumptions 100–101, 138–139
  basic model for 100–101, 141–143
  Hamiltonian for 145
 theory, semiclassical
  basic assumptions 100–102
  basic model for 100–101
  steady-state solution 123, 125
  threshold of oscillation 94, 96, 97, 107, 119, 127, 128, 230
  transition 91
Lateral mode 92
Lifetime
 for homogeneous broadening 171, 190, 201
 for inhomogeneous broadening 171, 188, 190, 208
 for spontaneous emission 97, 108, 120, 123, 159
 natural 3, 91, 94
Linear medium 106, 119
Linewidth
 natural 93, 95
 of laser field 99, 137, 157–159, 166
 of superradiant pulse 182
Local commutativity 31
Lorentz dispersion formula 119
Lorentzian
 lineshape factor 95, 117, 120, 127, 194
 spectrum 9, 159

Magnetic quantum number 179
Markoff process 167, 169
Maser 89, 99
Master equation 150, 235
Maxwellian velocity distribution 95, 133
McCall-Hahn equations 189, 191
Mean field theory 230
Michelson
 interferometer 2, 8–9, 82
 stellar interferometer 12
Mode
 axial 92, 103, 143
 expansion 38, 41, 74, 103
 lateral 92
 lifetime 91, 92, 97
 width 91, 93
Monochromaticity and second-order co-

herence 38
Multitime averages 169
Mutual coherence function 6, 10, 11, 82

Natural lifetime 3, 91, 94
Natural linewidth 93, 95
Negative absorption 107
Noise 137, 141, 159, 160, 167, 235
Noiselessness 37, 48
Nondiagonal representation 54, 68
Norm
 Hilbert-Schmidt 59, 61
 L^2 61
 trace-class 58
Normalized atomic lineshape factor 95 to 96
Normally-ordered
 characteristic function 62, 63, 66, 78, 84–87
 generating functional 78
 operator 55, 67
 product 27
Number operator 42
Nutation, optical 108

Off-diagonal
 equations of motion 116
 long-range order 39, 204
Onsager's regression hypothesis 169
Operator
 bounded 53, 58, 61, 69
 Hilbert-Schmidt 59–61
 isometric 58
 normally-ordered 55, 67
 positive 20, 32, 59, 67
 trace-class 58
 unbounded 53, 66
Optical "equivalence" theorem 56, 61 to 62, 63, 65–68
Optical resonator modes 91–93
Order
 of an entire function 51, 69, 86
 of coherence function 13, 28
Oscillator phase 44
Overcompleteness 49, 52, 213

Paley-Wiener theorem 57
Parametric conversion 207, 219
Partially coherent light 7

Pauli spin matrices 111–112, 173, 224
Pendulum analogy 180, 187, 189, 193, 202, 232
Permutation operators 179, 210
Phase
  curves 128–131
  diffusion 160–161, 166
  portrait 128
  reconstruction problem 10
  space 3, 49, 67, 128, 139, 212
  transition
    laser 204, 230
    superradiant 229–230
Photoelectric
  counting rate 21, 25
  detection, theory of 20–25
  detector, ideal 20–22
Photoelectron statistics 78, 156
Photon
  avalanche 89
  bunching 80–81, 171, 183
  echoes 171, 184–187, 208
  number
    and phase 44
    average 44, 65, 75, 86
    fluctuations in 76–77
  statistics 137, 143, 149, 222, 231
    for coherent state 75, 80–81, 155
    for laser field, 151–156
    for thermal light 71, 76, 80–81, 153
Picosecond pulses 207
Picture
  Heisenberg 43, 139, 167
  interaction 22, 145, 158, 164
  Schrödinger 22, 143, 145, 158, 164
Planck's law 71
Poisson distribution 44, 75–78, 156, 223
Polarization
  macroscopic 102, 113, 189, 190, 217, 230, 232
  microscopic 102, 107
Population
  densities 116, 122
  inversion 90, 96, 107, 116, 133
  inversion density,
    average 118, 120
    critical 94, 97, 117, 119, 125, 152, 230
    zero-field 123
Position and momentum operators 43, 46

Positive-definiteness conditions 32–34
Positive-frequency part 4, 18, 23, 26, 104
Positive operator 20, 32, 59, 67
Power spectrum 9
Precession 112–113, 173, 179, 181, 185
Probability conservation 112, 190
Product states 175, 217
Pseudo-electric dipole moment 173, 179, 185, 189, 195, 197
Pseudo-spin 176, 212
Pulling effect 107, 120, 132, 188
Pulsation effects 122
Pulse
  area 193, 197
  break-up 199–200, 202, 207
  compression 200, 207
  delay 200, 207
  $\pi/2$ 185
  π 186, 201
  2π hyperbolic secant 185, 193, 194, 203
  self-transparent 191
  superradiant 180–182, 187
  velocity 188, 194, 196, 202
Pumping 90, 96
  parameter 165
  rate 114–115, 122, 133, 141
  reservoirs 101, 115, 122, 127, 204

"Q" factor 92–93, 97, 106, 121, 149, 159
Quantization of radiation field 17–18, 143, 218
Quantum theory of the laser, see Laser theory, quantum
Quasi-monochromatic light 5, 7, 10, 11
Quasi-particle excitations 226, 229
Quasi-periodicity 232
Quasi-probability distribution 68, 164, 214
Quenching of oscillation 129

Rabi flipping frequency 208
Radiation
  gauge 17
  rate
    from arbitrary state 177
    from coherent atomic state 216
    from Dicke state 177, 216, 231, 233 to 234
Radiationless states 178

Raising and lowering operators 144, 175
Rate equations 122
Reconstruction
 phase 10
 of coherence functions 34, 84–87
 theorem 86
Reduced density operator 39, 139–141,
 143, 146, 203
Refractive index 106–107, 119
Regression of fluctuations 168–169
Relative excitation 125
Representation
 diagonal, see Diagonal coherent-state re-
 presentation
 Fock, see Fock representation
 Weyl 60, 68
 Wigner 68
Reservoirs, gain and loss 101, 122, 127,
 139, 141, 161, 231
Residual interaction 227, 228
Reslution of unity 50, 161, 214
Rise time of laser field 119, 138, 157
Rotating frame 184–185, 189
Rotating-wave approximation 100, 104,
 118, 121–122, 146, 159, 167, 215,
 218

Saturation
 effect 98, 117, 123, 127, 153, 164, 200
 parameter 127, 153, 164
 rate 123–124
Schrödinger picture, see Picture, Schrödin-
 ger
Schwarz's inequality 7, 59
Schwinger's theory of angular momentum
 219
Second-order coherence functions 6, 13,
 21
Selection rules 175, 177, 182, 183
Self-consistent field 102, 230
Self-focussing 200, 207
Self-induced transparency 171, 187, 200
 to 201
Self-stimulation 220
Self-transparent pulse 191, 193–194
Semiclassical
 radiation theory 99
 theory of the laser, see Laser theory,
 semiclassical

Separable Hilbert space 224, 225
Shock wave 201
Slowly-varying amplitude and phase approximation 104, 188, 190, 196
 solitons 203–204
Space
 $\mathscr{D}$ 57, 61
 $\mathscr{D}'$ 57
 $\mathscr{S}$ 61
 $\mathscr{T}$ 57
 $\mathscr{T}'$ 57
Spectral
 condition 31
 density 9, 34
Spectrum, Lorentzian 9, 159
Sphere, Bloch 179, 186, 195, 212, 214
Spin
 echoes 184
 flips 226
Spontaneous emission 28, 31, 91, 94, 99,
 108, 137, 141, 218, 220, 232, 235
 lifetime 97, 108, 120, 123, 159
 rate 174, 177, 233
Stability of laser modes 128–131
State
 coherent atomic, see Coherent atomic
 states
 coherent, see Coherent states
 Dicke, see Dicke states
 Fock, see Fock states
 fully excited 191, 201, 217, 231–232,
 235
 mixed 18, 21, 83
 product 175, 217
 pure 18, 19, 21, 83
 superradiant coherent atomic 180, 217
 superradiant Dicke 177–178, 220, 231
 to 232, 235
Stationary fields 6, 9, 14, 30, 33, 34, 72,
 78, 79, 83
Stationary states for Dicke's model 220
 to 223
Statistically homogeneous field 30, 33
Statistically isotropic field 30
Stellar interferometry 10–12
Stereographic projection 213
Stimulated
 Brillouin scattering 219
 emission 27–28, 89, 220

Stimulated
   see also Induced emission
   Raman scattering 219
$SU(2)$ group 222
Superconductivity and coherence 39
Superfluidity and coherence 39, 204–205
Superradiance 171, 177
   classical model of 179
   two-photon 209
Superradiant
   coherent atomic state 180, 217
   Dicke state 177–178, 220, 231–232, 235
   ground state 229
   phase transition 229–230
   pulse 180–182
Susceptibility, complex 106

Telegraph equation 103
Thermal light,
   coherence functions for 87
   coherence time 3
   degeneracy parameter 4
   density operator
      in coherent-state basis 72
      in Fock basis 71
   photon statistics for 71, 76, 80–81, 164
   see also Blackbody radiation
Thermal noise 137, 160
Thermodynamic limit 223, 224
Threshold, *see* Laser threshold
Time-ordered product 25
Tipping angle 193, 197, 216
Titchmarsh's theorem 5
Trace 19
Trace-class
   norm 58
   operator 58
Transient
   behavior of laser 156–157, 166
   interference effects 82–84
   optical nutation 208
Transition
   dipole moment 23, 109, 114, 117, 168, 172–173, 181, 208
   frequency 108, 121
Transmissivity 92
Tridiagonal matrix 222, 229
Twisted convolution product 61
Two-level atoms 100

   equations of motion for 107–111
   geometrical treatment 111–113
Two-photon superradiance 209
Type of entire function 86

Ultradistribution 57
Unbounded operator 53, 66
Uncertainty principle 48, 67, 117, 212, 218

Vacuum expectation values 29
Vacuum fluctuations 99
Van Cittert-Zernike theorem 10
Van der Pol
   equation 127, 131, 168
   oscillator 131
Velocity
   of pulse propagation 188, 194, 196, 202
   phase 107, 188
Visibility of interference fringes 8, 10, 12, 34

Wave equations for mutual coherence 10
Wave packet
   Gaussian 47
   minimum-uncertainty 48
Weak equivalence class 226
Weight function of diagonal representation
   and antinormally-ordered density operator 161–162
   and normally-ordered characteristic function 63, 85
   for chaotic field 72
   for ideal laser light 74
   for single mode 54, 62, 63
   integral equation for 55
Weisskopf-Wigner approximation 108, 141–142, 150
Weyl
   representation 60, 68
   rule of association 68
White noise spectrum 167
Wiener-Khintchine theorem 9, 160–161, 167–168
Wightman functions 29, 203
Wightman's reconstruction theorem 34
Wigner representation 68

Young's interference experiment 3, 6, 31, 34, 39

Zero-point fluctuations 43, 48, 137, 144